CAMBRIDGE LIBRARY COLLECTION

Books of enduring scholarly value

Zoology

Until the nineteenth century, the investigation of natural phenomena, plants and animals was considered either the preserve of elite scholars or a pastime for the leisured upper classes. As increasing academic rigour and systematisation was brought to the study of 'natural history', its sub-disciplines were adopted into university curricula, and learned societies (such as the London Zoological Society, founded in 1826) were established to support research in these areas. These developments are reflected in the books reissued in this series, which describe the anatomy and characteristics of animals ranging from invertebrates to polar bears, fish to birds, in habitats from Arctic North America to the tropical forests of Malaysia. By the middle of the nineteenth century, this work and developments in research on fossils had resulted in the formulation of the theory of evolution.

Taxidermy

First published in 1840 as a volume in the *Cabinet Cyclopaedia* – a series published between 1830 and 1844, intended for the self-educating middle class – this work was written by the naturalist and artist William Swainson (1789–1855). The first part is a treatise on taxidermy, showcasing methods of Victorian science that may appear gruesome to modern readers. It discusses the best ways to collect, preserve and present animals for scientific study. Swainson gives detailed advice, making allowances for naturalists working in different locations and searching for a range of species. The directions for skinning and mounting animals are not for the faint-hearted, but they offer a fascinating insight into the practices of the time. The work's second part is a zoological bibliography, with short biographies of notable authors. Zoological painters and engravers, such as Thomas Bewick (1753–1828), are also featured.

Cambridge University Press has long been a pioneer in the reissuing of out-of-print titles from its own backlist, producing digital reprints of books that are still sought after by scholars and students but could not be reprinted economically using traditional technology. The Cambridge Library Collection extends this activity to a wider range of books which are still of importance to researchers and professionals, either for the source material they contain, or as landmarks in the history of their academic discipline.

Drawing from the world-renowned collections in the Cambridge University Library and other partner libraries, and guided by the advice of experts in each subject area, Cambridge University Press is using state-of-the-art scanning machines in its own Printing House to capture the content of each book selected for inclusion. The files are processed to give a consistently clear, crisp image, and the books finished to the high quality standard for which the Press is recognised around the world. The latest print-on-demand technology ensures that the books will remain available indefinitely, and that orders for single or multiple copies can quickly be supplied.

The Cambridge Library Collection brings back to life books of enduring scholarly value (including out-of-copyright works originally issued by other publishers) across a wide range of disciplines in the humanities and social sciences and in science and technology.

Taxidermy

With the Biography of Zoologists

WILLIAM SWAINSON

CAMBRIDGE
UNIVERSITY PRESS

CAMBRIDGE
UNIVERSITY PRESS

University Printing House, Cambridge, CB2 8BS, United Kingdom

Published in the United States of America by Cambridge University Press, New York

Cambridge University Press is part of the University of Cambridge.

It furthers the University's mission by disseminating knowledge in the pursuit of
education, learning and research at the highest international levels of excellence.

www.cambridge.org
Information on this title: www.cambridge.org/9781108067775

© in this compilation Cambridge University Press 2014

This edition first published 1840
This digitally printed version 2014

ISBN 978-1-108-06777-5 Paperback

THE

CABINET CYCLOPÆDIA.

CONDUCTED BY THE

REV. DIONYSIUS LARDNER, LL.D. F.R.S. L.& E.

M.R.I.A. F.R.A.S. F.L.S. F.Z.S. Hon. F.C.P.S. &c. &c.

ASSISTED BY

EMINENT LITERARY AND SCIENTIFIC MEN.

Natural History.

TAXIDERMY,

BIBLIOGRAPHY, AND BIOGRAPHY.

BY

WILLIAM SWAINSON, A.C.G. F.R.S. & L.S.

HON. F.C.P.S. ETC., AND OF SEVERAL FOREIGN SOCIETIES.

LONDON:

PRINTED FOR

LONGMAN, ORME, BROWN, GREEN, & LONGMANS,

PATERNOSTER-ROW;

AND JOHN TAYLOR,

UPPER GOWER STREET.

1840.

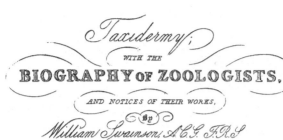

Taxidermy;

WITH THE

BIOGRAPHY OF ZOOLOGISTS,

AND NOTICES OF THEIR WORKS,

By

William Swainson, A.C.G. F.R.S

&c. &c.

E. Finden

William Swainson

London:
PRINTED FOR LONGMAN, ORME, BROWN, GREEN & LONGMANS, PATERNOSTER ROW;
AND JOHN TAYLOR, UPPER GOWER STREET.
1840.

THE

CABINET

OF

NATURAL HISTORY,

CONDUCTED BY THE

REV. DIONYSIUS LARDNER, LL.D. F.R.S. L.&E.

M.R.I.A. F.R.A.S. F.L.S. F.Z.S. Hon. F.C.P.S. &c. &c.

ASSISTED BY

EMINENT SCIENTIFIC MEN.

TAXIDERMY,

BIBLIOGRAPHY, AND BIOGRAPHY.

BY

WILLIAM SWAINSON, A.C.G. F.R.S. & L.S.

HON. F.C.P.S. ETC., AND OF SEVERAL FOREIGN SOCIETIES.

LONDON:

PRINTED FOR

LONGMAN, ORME, BROWN, GREEN, & LONGMANS,

PATERNOSTER-ROW;

AND JOHN TAYLOR,

UPPER GOWER STREET.

1840.

London:
Printed by A. Spottiswoode,
New-Street-Square.

CONTENTS.

PART I.

A TREATISE ON TAXIDERMY.

PART II.

THE BIBLIOGRAPHY OF ZOOLOGY,

CONTENTS

PART I.

A TREATISE ON TAXIDERMY.

CHAPTER I.

ON COLLECTING ZOOLOGICAL SUBJECTS.

(1.) THE economy of animals can only be studied
when the functions of life are in full activity; their
haunts must be explored, their operations watched, and
their peculiarities observed in the open air. But in
order to acquire a more accurate knowledge of their
external form, and to investigate their internal structure,
it is absolutely necessary to examine them in a dead
state. Hence has arisen the art of TAXIDERMY, which
teaches the various processes by which the form and
substance of animal bodies may be preserved from decay,
and rendered subservient to the studies of the naturalist in his closet. It is an art, therefore, absolutely
essential to be known to every naturalist; since, without
it, he cannot pursue his studies or preserve his own
materials. As such, our present treatise forms an essential, although, perhaps, a subordinate, part of the
" CABINET OF NATURAL HISTORY." We shall here
consider taxidermy in its most extended sense, under
the several heads of—1. Collecting, 2. Preserving, and
3. Arranging, animal productions.

(2.) It is not necessary that a zoological collector
should be a scientific naturalist; or that he should un-

derstand any other than the practical or mechanical parts of the science. Nevertheless, many subordinate qualifications should be possessed by those who follow this occupation in foreign climates. Strength and activity of body, a quick and discriminating eye, capable of perceiving at once minute distinctions; with a courageous, persevering, and inquiring spirit, — are all necessary to insure success : to these should be added a general acquaintance with the elementary principles of zoology, and a complete knowledge of taxidermy in all its branches. The methods of collecting the different tribes of animals, and the apparatus necessary to be provided, are so various, that we shall hereafter give a detailed description of both, under the classes to which they are more immediately applicable. We shall treat the present subject, in fact, somewhat in a professional way; chiefly addressing ourselves to those who collect in foreign countries, either for themselves or others.

(3.) The general equipment of a travelling naturalist or collector should consist more or less of the following articles : — A double and single barrel gun, with an ample assortment of caps, flints, shot, spare screws, &c. If he is proceeding to Africa or India, where the larger quadrupeds are found, a rifle will be advantageously substituted for a single barrel gun. Dissecting instruments for opening quadrupeds, birds, &c. Preserving drugs and preparations. Bottles for containing subjects in spirits, fitted into cases. Canvass knapsacks. Corked store boxes for insects, and others for the pocket and for immediate use. Pins of all sizes. Boxes fitted with moveable trays for bird skins. Apparatus for collecting insects. Chip boxes of different sizes, for small and delicate shells, &c. Knives, scissors, needles, thread, &c.— As a general rule, the collector proceeding abroad should adapt the size of all his packages to mule or horse carriage : such are, indeed, the only conveyances he will find throughout South, and over a large portion of North, America. In Southern Africa, wagons are used on long journeys, but on short ones the baggage is conveyed upon the backs either of horses or oxen. The naturalist who collects in Europe and other countries more civilised, will of course require a much more scanty equipment.

(4.) QUADRUPEDS.—The best information respecting the species of quadrupeds inhabiting any particular district, and of their local haunts, can readily be obtained from the natives, whose assistance may be called in with advantage, and secured by a competent reward. Gentlemen resident abroad, particularly in India, have it very much in their power to benefit the museums of this country, by acquiring a sufficient knowledge of taxidermy to enable them to preserve the skins of animals killed in the chase, since the public and private collections in Britain are very deficient in many of these species. The skulls and horns, where it is inconvenient to preserve the entire skin, are objects of much interest, particularly if accompanied with drawings, measurements, and notes of the habit, food, &c. The collector in Southern Africa should pay particular attention to the different species of antelopes and rhinoceros : the former are very numerous ; and of the latter, the skulls and horns will be sufficient to identify the species. We are still ignorant of many quadrupeds of the north-west coast of America, whose furs are articles of commerce. The skins may be removed and slightly stuffed in the manner hereafter described. If the skulls alone are desired, they may be easily prepared by parboiling the head, and separating the fleshy parts with a knife or scraper : the brain is either removed through the occipital hole, or (if the animal is large) by sawing the skull in two ; when clean, and the smell evaporated, the parts are tied together and left to dry in the shade. The *packing* of the skins or bones of quadrupeds requires but little care ; they must, however, be well dried before they are put into the case, the sides and joints of which should be perfectly close and waterproof.

(5.) Living quadrupeds have long been brought to this country as articles of commercial speculation ; but they are now likely to be imported for the purposes of science. The younger they are procured, the less difficulty will be found in rearing and in reconciling them to confinement. When first taken, great attention should be paid to their food, which should assimilate as much as possible with that which they prefer in a wild state. Every effort should be made, if not to tame them, at least to reconcile them to the presence of strangers and other persons besides their keeper. This is not difficult, as there is scarcely any animal whose ferocity cannot be softened by kind and judicious treatment. An excess of food is at all times bad ; and the greatest attention to cleanliness is equally essential. Animals intended to be sent by sea, should be put into confinement two or three weeks before the ship sails, that the change may not appear to them so great ; and the passengers should be particularly requested not to irritate or worry them during the voyage.

(6.) BIRDS.—A collector of birds should be provided with one or two light fowling pieces, and duplicate parts of all their usual apparatus ; a supply of the best powder contained in tin canisters, and of shot in bags : he may take with him a small quantity of swan and duck shot, but he will find Nos. 6. and 8. the most useful ; while small birds not larger than a sparrow are killed with the least injury to their plumage by what is called dust shot. For preserving his specimens, he must have a good supply of arsenical soap, penknives, sharp and blunt pointed scissars, &c. Cotton or tow can be had in America, and in some parts of India; but if the collector proceeds to Africa or the South Seas, it will be prudent to take a small stock with him. The best periods of the day for procuring birds are early in the morning and late in the evening. In warm latitudes, the sportsman should always choose the dawn of day for his excursions, not only on account of the refreshing coolness of the air, but as being that time when the greatest number of birds are seen and heard. A little boy can carry the box or basket intended to hold the game ; and this will enable the sportsman to enter the thickets and woods with less difficulty : the box should contain some pieces of soft paper, and a little cotton or tow. Before each bird is put in, the feathers are smoothed, and a small piece of cotton or tow twisted round the bill and nostrils. If the wound bleeds much, some tow should be laid upon it, and the bird wrapped in paper, to prevent it from soiling the others : the box should be made of tin, as this metal keeps the specimens cool. In two or three hours a suffi- cient number may be killed to occupy the collector during the rest of the day in stuffing : by that time, or about eight or nine o'clock, the great heat commences, the birds become silent and retire to the deep shades, and the sports- man had better return home. Towards the cool of the evening the birds again emerge from the woods, and, if any more subjects are wanted, the sportsman may again use his gun. In these climates, birds will not keep be- yond a day without some degree of putrefaction taking place : this shows itself by the feathers coming off ; first on the belly, and after on the front : it is, therefore, ad- visable not to shoot more specimens than can be prepared in twenty-four hours. Birds in tropical countries are, in general, so tame, that they can be approached very near ; there is, therefore, little occasion to be very par- ticular about the excellency of the gun or the quality of the powder : the first, for convenience, should be light, and the last good. Humming-birds are advantageously shot when hovering over the flowers on the nectar of which they feed ; but the charge should be very small, and dust shot alone used. Birds of the size of a hawk or thrush may be killed with shot No. 8. In some parts of America, the natives shoot the creepers and humming- birds with a blow pipe. An expert marksman of this sort might be retained in the service of the collector, as the specimens are killed without the least injury to their plumage, and consequently in the best state for preserv- ing.* The sexes of every species should be industriously

* *The Naturalist's Guide*, for collecting and preserving all Subjects of Na- tural History and Botany ; intended for the use of students and travellers. By W. Swainson. With plates. London : second edition, 12mo. p. 18.

sought after, and no pains should be spared in watching their manners and habits. Their nest, if remarkable in form or construction, may likewise be procured. When the skin has been slightly filled with cotton, and sewed up, it may be laid with others, upon any soft substance, within one of the trays of the travelling box, hereafter described, and suffered to dry: each specimen should be numbered, or have a label attached to it, specifying the sex, the place and date when found, the contents of its stomach, and any other particulars that may be known or observed relative to its habits or economy. When dry, the skins may be enveloped in cotton or paper, and closely packed in the trays of the box. Unless the boxes are rendered air-tight, by the seams being pitched, it would be prudent to inspect them every week or ten days, until they are finally sent on board. Birds of England, and other parts of Europe, are collected without much difficulty, and by means well known. It may, however, be observed, that, in our northern temperature, specimens may be sent from one end of England to the other in a fit state for preservation (except during the height of summer), by packing them in a close tin box, partially filled with powdered charcoal.

(7.) Skins of birds should be packed in well-made boxes ; each specimen wrapped either in cotton or tow, or in paper, and the interstices filled with moss or any other soft substance. When the lid is shut and secured, the joints and seams should be closed with tow and pitched over. Small boxes may be papered ; but a little arsenic or corrosive sublimate should be mixed in the paste, otherwise it will, in tropical climates, be attacked and fed upon by ants.

(8.) REPTILES and SERPENTS are best procured by the natives. Indeed, the danger that would result from a *serpent* hunt is too great to warrant the collector setting out on an excursion of that sort. The different species are generally well known to the country people, who give them provincial names, and who may safely be consulted respecting their habits. Every information on these points the collector will be careful to note among his memorandums.

(9.) FISH.—During a voyage, many species may be caught by a hook and line thrown over the stern of the vessel: these should be drawn and described, if too large for preserving ; the lesser species can be put into bottles of spirits, a few of which should always be in readiness. The most advantageous modes for collecting fish are, either to accompany the fishermen in their boats, to be present at the drawing of their nets, or to frequent the markets at an early hour : all these plans, indeed, should be followed, as those species only are exposed for sale which are considered good for food ; and consequently many others, particularly those of a small size, are thrown back into the sea. Fishing-nets, at the same time, bring up many other marine animals, as crabs, corals, starfish, medusæ, &c. Those which the collector may select, should be put in a bucket of salt water, that their movements and different organs may be examined, and if possible drawn and described. Fishermen's boys may likewise be impressed into the service, and instructed to throw into a bucket the refuse of the nets. All these plans we pursued with much success, both in the Medeterranean and in the ports of America. A few plain hooks and lines is all the equipment that the travelling naturalist need require in this department, independent of the boxes which contain the bottles for receiving the specimens, unless he is provided with a small trawl or casting net, either or both of which would be very desirable if his plans are likely to place him on the sea coast. In England, the western shores of Devon and Cornwall are frequented by a great variety of species, of the most beautiful forms and colours ; and on all these shores the ichthyologist will obtain an abundant harvest. So productive is the west of England in marine animals, that the late active and intelligent colonel Montagu, although living upon the spot for many years, was continually adding new subjects to his collections.

(10.) Insects, to be collected with success, require a knowledge of their haunts, great quickness of hand, and accuracy of eye. The apparatus of an entomologist, particularly if he is proceeding to a warm country, requires to be minutely described. The different instruments used in catching insects are as follows : — The fly-net, elastic net, bag-net, hoop-net, landing-net, forceps, and digger. For securing insects alive, are required phials, chip boxes, and breeding cages; and for preserving them when dead, pins, braces, pocket boxes, store boxes, and travelling chests.

(11.) *The fly-net* (*fig.* 1.) is preferred above every other by English collectors, and perhaps with reason, as 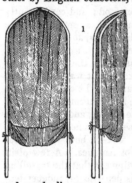 it may be applied to more purposes than one. This instrument is similar in its construction to a bat-fowling net, and is more particularly adapted for capturing flying insects. The net itself should be made of strong green gauze : a white colour is preferred by some ; but this is objectionable for night flying insects. The rods are from 5 to 6 feet long, half an inch diameter at the base, and gradually tapering towards the end. We prefer those made of hazel wood, from their lightness, although they are usually formed from ash. Each rod may consist of three straight pieces or joints, besides the last or curved one, which is generally made of cane, and thus yields to any sudden knock or pressure : each of these joints at one end has a simple brass tube or ferrule, which receives the bottom of the adjoining piece, like a fishing-rod : the end which goes into the ferrule should have a notch or check, to prevent it from twisting. The terminal joint, being of cane, can either be bent into a curve or fitted into an angular ferrule, so

as to form an obtuse angle with the rest of the rod ; but
the former is the better plan. This frame-work of the
net may, for convenience of travelling, be contained in
a canvass bag, and carried in a long pocket made in the
vest. After the gauze is cut into shape (it must always
be loose), it is welted round and bound with a strong
ribbon, that the rods may pass within : in the centre of
the upper part, where the tops of the two rods meet, a
piece of leather is sewn across, which thus forms a sort
of hinge : at the other extremity the gauze is folded,
so as to form a bag, by which the escape of captured
insects is prevented. Finally, that the net may be
securely fixed upon the rods, two strings are sewed at
the bottom of each side, which pass through holes made
in the rods, about six inches from their handles : without this security, the net would be perpetually slipping
upwards. The manner of using this kind of net for
catching flying insects, is to hold it in both hands, in
an extended position ; and so soon as you have brought
it fairly beyond the insect you are pursuing, suddenly
closing it, at the same time giving it a slight jerk upwards. Insects resting upon the ground, may be captured by quickly spreading the net over them, and then
closing it. It may be likewise advantageously employed
to receive insects beaten from bushes and trees, by
either holding both the sticks in one hand and beating
the boughs with the other, or by extending it upon the
grass.

(12.) Maclean's *elastic net* is thus described by
Mr. Kirby : — " It is constructed of two pieces of stout
split cane, connected by a joint at each end and with a
rod which lies between them in which a pulley is fixed:
through this a cord fastened to the canes passes : a long
cane with a ferrule receives the lower end of the rod,
and forms a handle ; and to the canes is fastened a net
of green gauze. Taking the handle in your right
hand and the string in your left, when you pull the
latter the canes bend till they form a hoop, and the net
appended to them is open ; when your prey is in it,

relax the cord, and the canes become straight and close
the mouth of the net: keeping them close with your
left hand, you may soon disable your prey with your
right. This net was invented by Dr. Maclean, of Col_
chester, who has scarcely ever found it to fail." *

(13.) *The bag-net* (*fig.* 2.), invented by Mr. Paul, of
Starston, Norfolk (called by many the turnip-net), is

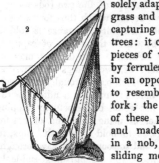

solely adapted for brushing the
grass and other plants, and for
capturing insects beat from
trees: it consists of two stout
pieces of wood fixed obliquely
by ferrules into a handle, each
in an opposite direction, so as
to resemble a wide-spreading
fork ; the other end of each
of these pieces is left thick,
and made to curve upwards
in a nob, for the purpose of
sliding more readily over the
ground : towards this end they are united by a cross
bar of strong iron wire: the circumference is thus
made triangular, and to it is fastened a long, loose
bag, gradually tapering towards the end, and made
of glazed calico. To use it for brushing herbage, the
collector shoves it before him through the fields. This
may be done with one hand, while the other can be occupied in beating any insects that may be on the taller
shrubs, into the net with a stick: from time to time the
contents are either examined, or shaken into the bottom
of the bag, which, being very narrow, sufficiently con-
fines them. This net is well adapted for certain descriptions of insects; but, from its nature, cannot be
used for such as are flying, or in conjunction with any
other. The size is not material, but the handle should
be sufficiently long to enable the collector to use it without stooping.

(14.) *The hoop-net* is more in use among the Con-

* Kirby and Spence, iv. p. 520.

tinental collectors than any other: it consists of a hoop about 10 or 12 inches in diameter, made either of strong brass or iron wire, with a socket or ferrule to allow of its being fastened to the end of a long stick or pole: the hoop supports a bag of fine gauze or muslin, about 20 inches deep.   For the convenience of carriage, and for occasional rambles, the frame-work or hoop may be made of two semicircular pieces, united by joints, and capable of being folded for the pocket.   To use it with adroitness requires some practice: the collector aims at a flying insect; and at the moment he fancies it is within the circumference of the hoop, he gives it a sudden and abrupt twist, accompanied by a forward jerk, so as to close the opening and secure the insect within.   The pole or stick should be tapering, and be rendered capable of being lengthened by the addition of joints.   Entomologists who are proceeding abroad, however much they may prefer the English fly-net, should nevertheless provide themselves with two or three bag-nets to keep in reserve: the first, being more complex, is liable to get out of repair, or to break; but the bag-net is particularly simple; and the stick, or handle, can be procured wherever trees or canes are found.

(15.) *The landing net* is on a similar construction to the last; but the hoop is made much stronger, and a net, with very small meshes, is substituted for the gauze. This net is used only for capturing aquatic insects by the entomologist, but is of great service to the conchologist in bringing up small freshwater shells.   As many of the aquatic insects are very minute, the net should be lined inside with strong muslin, sufficiently coarse to admit the escape of the water.   The bag-net may be used for the same purpose; but the gauze will be liable to tear, and it cannot be employed to catch terrestrial insects again until it is clean and dry.

(16.) *The forceps* (*fig.* 3.) is an essential instrument to every collector, who should always have two or three of different sizes in reserve in case of accidents: it is made of iron; the handles very much resembling those

of a pair of scissors, or, as Mr. Kirby observes, an old pair
of curling irons might be made into very good handles.
The hoops or leaves of the forceps may be either round or
octagonal, and should be made of brass, to prevent rust ;

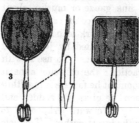

5 or 6 inches in diameter
will be a very good size ;
the joint of the handle
should be placed nearer
the rings for the thumb
and finger than to the
leaves, or the instrument
will not open well. Green
gauze or transparent mus-
lin is fastened tight round
the frame, although some collectors prefer catgut for
this purpose : over the gauze, the rims should be bound
with a narrow strip of thin leather, otherwise they
will require frequent covering. This instrument is particularly useful for capturing winged and other insects
when at rest, more especially *Diptera* and *Hymenoptera:*
it is used by sufficiently extending the leaves, measuring
the distance with the eye, and then suddenly closing
them upon the insect. The handles are sometimes made
10 or 12 inches long ; but the general length is from 5
to 6 inches. The forceps may be carried without inconvenience in a side pocket of the vest, and should
accompany the collector in all his rambles. Such persons as intend carrying any of these instruments abroad
will do well to provide themselves with a piece of spare
gauze, or muslin, to re-cover them when they become
injured.

(17.) *The digger* (*fig.* 4.) is an iron instrument, very useful for raising the bark and
digging at the roots of trees for insects to be
found in such situations : it is about 6 inches
long, fitted into a wooden handle, the point
being arrow-shaped. A strong knife may be
employed for the same purpose.

(18.) The insects being captured, there will be some which the collector may wish to preserve alive, either to watch their transformation, or until his return home. For small coleopterous insects, a phial or tin box is very useful. The aperture of either must be small, and furnished with a tube extending one or two inches down the neck, to prevent the insects from crawling out; a chip wafer-box, having a small hole fitted with a cork, may likewise be used. Minute insects may be secured in quills, having one of their ends stopped with cement, and the other fitted with a cork stopper. Chip boxes of various sorts and sizes are likewise essential for securing living specimens. Pill-boxes are chiefly used for inclosing such of the smaller species of *Lepidoptera* as are more conveniently killed when brought home. The collector should be furnished, likewise, with a pocket larva box of convenient size, with a piece cut out of the top, and covered with gauze for the admission of air : it should be sufficiently filled with the leaves upon which the caterpillars feed, to prevent them from being shaken and injured.

(19.) *The breeding cage,* for the rearing of cater-pillars, is only useful to collectors whose residence for some considerable time is stationary. Mr. Kirby strongly recommends a box or cage invented by Mr. Stephens (one of the most celebrated British collectors of the present day), which he thus describes : — " The length of the box is 20 inches, height 12, and breadth 6 ; and it is divided into five compartments. Its lower half is con-structed entirely of wood, and the upper of coarse gauze, stretched upon wooden or wire frames : each compart-ment has a separate door, and is, moreover, furnished with a phial in the centre, for the purpose of containing water, in which the food is kept fresh ; and is half filled with a mixture of fine earth and the dust from the in-side of rotten trees ; the latter article being added for the purpose of rendering the former less binding upon the pupæ, as well as being highly important for the use of such larvæ as construct their cocoons of rotten wood. The chief advantages of a breeding cage of the above construction are, the occupation of less room than five separate cages, and a diminution of expense; both important considerations, when any person is engaged extensively in rearing insects. Whatever be the construction of the box, it is highly necessary that the larvæ be constantly supplied with fresh food, and that the earth at the bottom should be kept damp. To accomplish the latter object, I keep a thick layer of moss upon the surface, which I take out occasionally (perhaps once a week during hot weather, and once a fortnight or three weeks in winter), saturate completely with water, and return it to its place : this keeps up a sufficient supply of moisture, without allowing the earth to become too wet, which is equally injurious to the pupæ with too much aridity. By numbering the cells, and keeping a register corresponding with the numbers, the history of any particular larva or brood may be traced."* A breeding cage, described by Mr. Kirby, and well adapted for common purposes, may be formed of " a cubical box of moderate dimensions, glazed in front, or on one side, to enable you to watch their proceedings, with the other sides and top fitted with fine canvass for the admission'; or the box may be canvassed all round, with a door in front. In this you may place a small garden-pot filled with earth, with a phial of water plunged in it to receive the insect's food (*fig. 5.*). This might be moved when you wish to change the water without disturbing the earth, which should be kept somewhat moist." †

(20.) The breeding of exotic insects, more especially *Lepidoptera*, has been prosecuted to a great extent, and with important advantages to science, by a few zealous entomologists : among these, Mr. Abbott in Georgia,

---

* Kirby and Spence, iv. p. 540.   † Ibid.

North America, and Dr. Horsfield in Java, deserve particular praise. Few can hope to do what they have done in this department, without residing for years in one locality; but every exotic collector may have it in his power to breed a few insects, and thus supply some valuable information. There are thousands of exotic *Lepidoptera* familiar to our collections, but of whose metamorphosis we are entirely ignorant. We of course allude to the drawing or preservation of the larva and pupa, and of the plants upon which they feed. As this, to a person possessing a moderate proficiency in drawing, can be done in the house, and with very little trouble, we strongly recommend it to the attention of those who are residing in the East and West Indies, and in other of our colonies or garrisons. It requires no scientific study, and is both an exciting and an elegant amusement.

(21.) Other methods of capturing insects may be here shortly noticed. " The finger and thumb will be found a very handy forceps " on many occasions, and the fore-finger, slightly wetted, will best secure minute *Coleoptera* that are stationary or crawling upon any solid substance. A table-cloth, or other white piece of linen, spread upon the grass in the open breaks of woods, will allure many species. Carrion, dead birds, frogs, &c. will soon bring to them various insects, whose food consists of putrid bodies. Light is known to attract all insects, and it may be employed with the greatest advantage in tropical countries. In Brazil, during the rainy season, we placed the candles under glass shades (a lantern will answer the same purpose), opposite the open window, almost every evening. The number of insects that came into the room, attracted by the light, was sometimes truly surprising; it frequently happened, that, with all our apparatus laying upon the table, we could not capture them sufficiently fast to prevent many from escaping. Mr. Barrow[*], when travelling along the

* Travels in South Africa, i. p. 179.

banks of the Keiskamma river, in the month of September, observes, that near " fifty different species of moths came in one evening upon the table in my tent." " Entomologists," he continues, " collecting the *Phalænæ*, could not adopt a better plan than placing a tent with a light in it on the side of a wood." The annoyance which the Europeans experience in India from night-flying insects is well known; yet few of them, we apprehend, have found their way to European cabinets. Insects inhabiting excrement are mostly European: if the dung is immersed in water, the insects will soon leave it and mount to the surface, where they may be easily captured.

(22.) There are various modes of *killing* insects, according to their respective tribes. *Coleoptera* and *Hemiptera* should be brought home alive, in chip boxes or phials, and immersed either in spirits of wine or boiling water. Almost all other insects may be instantaneously killed at the time they are captured, by pressing the sides of the trunk, below the wings, with the finger and thumb. To do this expertly, and without injuring the specimen, requires some practice; the pressure must be regulated by the size and nature of the insect: young collectors generally spoil their early specimens by pressing them too much. Placing the heads of insects near the fumes of prussic acid is the only effectual mode of speedily killing large moths.

(23.) For *securing* insects when dead, a large assortment of pins is requisite; particularly for collectors proceeding abroad, where these necessary articles are seldom to be had. Some idea may be formed of the number required in tropical countries, by stating, that in Brazil we generally consumed one hundred in every day devoted to entomological pursuits. The most convenient size for the generality of insects are those not longer than eight tenths of an inch, and somewhat slender; but others, both above and below this standard. will be requisite for very large or very small insects. they may be stuck upon a pincushion suspended round the neck or at the button-hole.   Pocket store boxes are intended to receive the insects when first captured : those of an oval or oblong shape (generally used to hold toys) are mostly preferred ; they may be strongly papered outside, to exclude air and wet, and lined with cork at top and bottom.   The collector should be provided with several of different sizes, to suit the nature or extent of his hunting.   Mr. Kirby recommends " a mahogany one, about $7\frac{1}{2}$ inches by $4\frac{1}{2}$, and $1\frac{1}{4}$ deep in the clear, corked only at the bottom, and opening by pressing a spring, which can be done with one hand : this will contain as many of those insects that are generally transfixed with pins, as will be usually taken in a day's excursion."   Far otherwise, however, will be the case, if the collector is hunting in the tropics : he will require a box about 13 inches by 8, to contain the captures of a day, and will frequently be compelled to bring home the remainder of his game stuck both on the inside and outside of his hat : a small box, in those countries, is very useful for setting apart small and very rare insects.   In England, the collector will require, on a journey, only a moderate sized store box to receive the collected fruits of his daily excursions : this should be made with small brass hinges, and furnished with a lock and key : servants are always too curious, and it is best to keep such things out of their way.

(24.) *Setting boards* are used for placing the legs and wings of insects in their natural position, before they become stiff.   They merely consist of a thin deal board of a convenient size, covered with soft cork about a quarter of an inch thick : at one or both ends a loop or brass ring may be attached, to hang the board on the wall during the day, and while the collector is in the room ; but it should never be left exposed in this manner during the night, or even in the evening : there are two or three species of moths which lay their eggs on the insects, and thus may be the means of infecting a whole drawer or box of specimens.   We lost some

c

hundreds in this way, last autumn, before we discovered the secret cause of their destruction. In tropical countries, where every house is filled with ants, the setting board must be suspended by a string, dipped in pitch or bird-lime, to prevent the passage of these destructive creatures; but unless it is taken down during the night, the cock-roaches will reach it by flying, and destroy every speci-men. We shall describe the method of setting insects when noticing their preservation.

(25.) Insect *chests* are essential to collectors about to reside in distant countries for any length of time. We shall here give a description of one upon a plan somewhat novel,—two of which constantly accompanied us in our travels. The case is 2 feet 5 inches long, 1 foot 3½ inches broad, and (when the lid is closed) 1 foot 2½ inches high in the full; it is made of ¾ inch deal, painted lead colour, and the corners secured with iron clasps : the lid falls into a rabbet, and is 2¼ inches deep in the full, — thus reducing the height of the box, when opened, to 1¾ inch ; the lid is corked, and by having a thin inner lid *flush* with the external sides, it forms a large store box of itself. The body of the case itself is divided into twelve equal transverse partitions, by narrow slips of wood let in perpendicu-larly : each of these partitions receives a box made of thin deal, ⅞ths in thickness, and corked,—the outside depth of each being 2⅜ inches* ; they open in the middle, so that each half is of the same depth ; they are rabbeted, and fastened by hooks and eyes. Each of these boxes will contain, upon an average, 200 specimens : a small brass ring is fixed in the centre of each, by which it is easily drawn out. The advantages attending a case made upon these principles are many. By its size, it is exactly adapted to be carried on one side of a horse or mule, so that two constitutes an equal balance ; an advantage which can only be duly appreciated by those

* This is neither too deep nor too shallow — at least, for those collectors who do not stick their specimens at the top of long pins,— a method we absolutely detest.

who have had experience in these matters.   Secondly, by the vertical position of the boxes, any insects that may get loose by the shaking of the case, immediately fall to the bottom without injuring the rest.   Thirdly, the great number of specimens that may be preserved in such a small space.   Fourthly, the power it affords of immediately securing the contents of one box from injurious insects, by pasting it all over with paper, and rendering it air-tight, the moment it is filled.   And lastly, the advantage it offers of giving the collector additional room, as he proceeds upon his journey, should he have an opportunity of sending a part of the inclosed boxes, already filled, to his head quarters.

(26.) As a substitute for *cork*, the traveller will find, upon inquiry, other descriptions of soft wood, both in Asia and America, which may on an emergency be applied to the same purpose.   The stem of the American aloe is even preferable to cork, when the collection is stationary.   But as it does not hold the pin with the same tenacity as cork, it is by no means so good for travelling boxes.   A wood, nearly similar, but of a more compact substance, is often used to send insects from Java.

(27.) The implements, &c. required by the English collector, will of course be fewer in number and kind than those we have just enumerated: from these he can make his selection, and he will find additional information on this subject in Samouelle's *Entomologist's Compendium*, Kirby and Spence's *Introduction to Entomology*, and our *Naturalist's Guide*.

(28.) In searching for insects generally, scarcely any locality should be overlooked.   The greatest variety will, of course, be found in those situations producing the thickest and most diversified vegetation.   Hence the skirts of woods and forests, in all countries, abound with an infinitely greater number than are to be met with in an open country.   Yet the species found in one spot will not occur in the other, and *vice versâ*.   Thickets, hedges, lanes, heaths, commons, sand-pits, meadows, ponds, moss, decayed animal and vegetable bodies, banks, and

even sandy shores, are each the separate habitations of
innumerable tribes, which can only be discovered in such
situations.   The insects of Tropical America, notwithstanding their immense number and variety, are, generally speaking, very local.   The interior of this continent, so far as it has been explored, is more or less
arid and sandy, partaking very much of the nature of
Southern Africa.   In both, the coast is belted, for several
hundred miles, by an immense forest of lofty timber
trees, growing with a luxuriance of vegetation seldom, if
ever, seen in the interior.   It is here that nine tenths of
the insects hitherto brought to Europe are found.   As
the traveller passes towards the interior, and leaves these
fertile regions, the plants become scanty, and he may
frequently travel a whole day without observing a dozen
insects.   From the observations of Mr. Burchell, the
same character is observable in Southern Africa.   The
entomologist will therefore do well to regulate his proceedings accordingly; for he will be much disappointed
if he expects the interior of these continents will furnish
him with the same abundance of insects as he may have
observed on the coast.   On his excursions, he will find
it necessary to engage a boy to carry a part of his equipment ; the rest may be contained in a knapsack or canvass bag thrown over his own shoulders.   As regards
his dress, it should be as thin and light as possible, on
account of the excessive heat ; shoes, or lace-boots, made
of unblacked leather, are desirable, for their ease ; while
his hunting-hat had better be of straw, with a very broad
brim.   This answers a double purpose ; it screens him
from the sun, and is, upon emergency, a very good substitute for an insect box.

(29.) In collecting SHELLS, the implements required
are few in number. The principal of these is a ladle or
spoon, made of tin or thin iron, 5 inches long and
$3\frac{1}{2}$ wide, with a rim about an inch in height: it
should have a short hollow handle, by which it may
be fixed to the end of a long walking stick ; the middle
should be perforated with holes, no larger than is suf-

ficient for the passage of water. This instrument is
very useful in fishing for small river shells, or for sift-
ing fine sand on the sea-shore. One or two strong knives
will be necessary for separating limpets, ear-shells, &c.
from the rocks. A hammer and chisel, for procuring
such as perforate; and small tin boxes and bags for containing the specimens. In searching for the larger freshwater bivalves, a landing net, with very small meshes,
is of great service, and it may be made to fit upon the
same stick as that which receives the spoon already de-
cribed.

(30.) Marine shells are the most numerous, and there
are few situations on the sea-coasts which do not produce some species. The lowest ebb of the tide is the
best time for searching for them. The rocks, corals,
and stones, which are then left exposed should be carefully examined for chitons, limpets, ear-shells, and other
adhesive tribes, which are fixed upon the surface, or
shelter themselves in the crevices : they are detached by
suddenly passing a knife between them and the substance
they are upon. Muscles, and other gregarious bivalves,
furnished with a byssus, likewise occur in such situations. Wherever the rock, mud, or sand is pierced with
round holes, the collector may be tolerably sure of finding
bivalves : they are procured either by breaking the rock
with a hammer, or digging deep into the sand or mud
with a spade. The little puddles of salt water, left by the
tide, are the habitations of many univalve shells; and
others will be found beneath loose stones and sea-weeds.
If any shells appear to have been recently cast up on
the beach, and are not broken, they may be collected ;
but such as have lain some time, exposed to the friction
of the waves and the heat of the sun, are scarcely worth
that trouble. After a gale of wind, or violent storm,
the shore should be immediately visited, as fine shells
are frequently to be met with : if the line of coast is
extensive, a few boys should be engaged to assist in the
search. This must be done quickly; for it not unfre-
quently happens, that the next flow of the tide takes away

every shell.*   Small islands and coral reefs, not exposed
to violent surfs, are generally very rich in shells, particularly in different species of *Spondylus,* tree-oysters
(*Dendrostreæ*), clams (*Tridacnæ*), winged muscles
(*Margaritæ*), and other adhesive or byssiferous bivalves.

(31.) The *trawl,* or dragging net, upon a productive
coast, will generally bring up a variety of living shellfish, as well as of other marine animals.   Whenever
dead or broken shells are drawn up with the sounding
line, or observed upon the beach, they afford an almost
certain indication of the coast being productive.   The
trawl should be tried in every direction, both in deep and
shallow water; and when once the shelly ground has
been discovered, the collector may calculate upon procuring a variety of species peculiar to such waters.
Shellfish of a carnivorous nature may be caught in lobster
pots, which they frequent for the purpose of feeding upon
the offal used as baits.   In the Mauritian islands, as we
have heard, it is a common practice to fish for olive and
harp shells with a line and hook baited with flesh:
this method, no doubt, might be employed with great
advantage on other productive shores.   The fish markets
in Catholic countries should be regularly visited, particularly during the season of Lent, when shellfish
constitute an important article of food to the inhabitants.
In the market of Naples, we have often seen fine specimens of *Cardium spinosum* and *aculeatum, Pectunculus
pilosus, Pecten Jacobea,* and *varia, Murex brandanus,*
and many other species of a smaller size, thus exposed
for sale, merely for the sake of the fish.   Trawling in
the bay would produce, without doubt, a still greater
number.   At Taranto, according to Swinburne and
Ulysses, the variety and abundance of shellfish is prodigious: the latter author enumerates 185 species, found
by himself at Taranto and Naples. †   Shells, also, are
procured by divers or pearl fishers in various parts of

* Naturalist's Guide, 2d edit. p. 45.
† Travels in Naples, 1795, p. 512.

India. We have been told that the magnificent collection of shells formed by the late Mr. Griffiths in the island of Sumatra, were nearly all procured in this manner. The sea, in the sheltered bays and coves of tropical climates, is at times so clear and transparent, that objects are distinguished at the depth of fifteen or twenty feet. The collector should avail himself of this, by using a small hand-net fastened to a pole, by which the bottom may be scraped.

(32.) The most productive coasts for shells are those of the continent and islands of the Indian ocean, from whence near one fourth of the exotic species usually seen in cabinets are brought. It may be taken as a general rule, that the shores of islands abound with more shells than those of continents. Ceylon, Amboyna, Sumatra, and Java have long been celebrated for their shells; but those from Borneo and New Guinea are very little known. The island of Timor may be called the paradise of conchologists; for it has frequently been averred, that no part of the world can be compared with it in the variety and profusion of its marine productions. The coasts of Australia are considered productive, yet not particularly so. From the Pacific Islands many beautiful and rare species have been obtained ; and numerous others, in all probability, remain to be discovered. It is singular, that while the eastern coasts of South America are particularly barren, the western shores are found to be plentifully inhabited by testaceous animals, more especially those of the cyclobranchian tribe, or chitons, numerous species of which, of late years, have been received from Chili.   In Britain, the West of England, affords nearly two thirds of all the marine species yet discovered. The coasts of Exmouth, Sandwich, and Weymouth are particularly productive; so likewise are those of Tenby, Barmouth, Hastings, &c.   In Ireland, Dr. Turton has explored Bantry Bay, and the celebrated silver strand of Portmarnock, in Dublin Bay, with great assiduity and singular success; while in Scotland, a considerable number of rare and interesting shells have been discovered in the Frith of Forth, by Captain Laskey, and accurately described by him in the *Transactions of the Wernerian Society*.

(33.) *Fluviatile shells* may be sought for in fresh-water lakes, ponds, rivers, streams, and ditches filled by brooks. The greatest number of the univalves occur at or near the surface, under the leaves of aquatic plants, among decayed vegetables, &c. The bivalves, on the other hand, as also the *Ampullariæ*, *Melaniæ*, and *Paludinæ*, among the univalves, are only to be found at the bottom, either among the pebbles, or partly imbedded in the sand or mud: the first are easily captured by the hand, or by the spoon already described; but the different species of *Cyclas*, *Unio*, *Anodon*, &c., from fixing themselves within the mud (very often two or three inches beneath the surface), can only be extracted by a strong semicircular landing net, somewhat resembling a drag in miniature—the curved portion being that to which the handle is attached, while the straight side is in front: this side, which comes in contact with the bottom, might be furnished with three or four iron prongs, like a rake, which would detach the shells from the mud; while the net, being drawn forward, would receive them. Many of the European fluviatile bivalves are minute, and can only be secured by a net with very small meshes. There are scarcely any situations in this country where fresh-water shells may not be found. The exotic species should particularly engage the attention of the collector. The great rivers and lakes of North America abound with a surprising number of these bivalves, many of which grow to a very large size and astonishing thickness. Although we are now well acquainted with those of North America, few, comparatively, have yet been brought from the tropical regions of that continent, — still fewer from Asia, and scarcely any from Africa. As no cause has been assigned for such a singular disparity, we may presume it is occasioned by the fresh waters of those regions not having been sufficiently examined.

(34.) *Land shells* occur in all countries, and are found

in various situations; as humid spots covered by herbage, rank grass, &c.; beneath the bark, or within the hollows of old trees, crevices of rocks, walls, bones, &c.   Early in the morning, during a damp sunless day, or after showers of rain, land mollusks may be found crawling on the leaves of plants, the stems of trees, &c.   The animals will sometimes live in a torpid state for one or two years after they have been removed from their native country : it is therefore highly desirable that this experiment should be tried with a few of each species ; packing them in moss, or loose vegetable earth, but in such a way that they may not be shaken during the voyage.

(35.) The animals of all shells may be killed with warm water, in which they should remain two or three hours. The water must not boil, otherwise the colours, in many cases, will be changed or injured.   Previous to removing the animal, the shells should be simply cleaned with water and a hard brush.   Spirits of salt, or other acids, on no account should be used : they are, indeed, employed to remove scurf, or any extraneous bodies that sometimes hide the beauty of the specimens ; but their application requires much skill, and will prove destructive in the hands of inexperienced persons.   When the shells, there-fore, have been cleaned with a brush, the dead animals can be removed with a stout pin, or the point of a knife : the latter will be necessary for cutting the two muscles, generally found in bivalves, and by which the valves are closed.   The animals of these shells are never dead until these muscles are relaxed, and the valves begin to gape.   During this operation, great care must be taken not to injure the teeth ; and it is desirable that the ligament should be preserved entire.   The operculum, or lid, which closes the mouth of univalves, should be carefully detached, wrapped in paper, and replaced within the aperture.   The shells may be left to drain upon a towel and board placed in the shade.   In tropical countries, the assistance of ants may be called in with advantage.   We have been assured by a collector, who brought a large collection from Panama, that he left the

removal of the animals entirely to these industrious little
insects.

(36.) In *packing* shells, the smaller and more delicate
kinds will be best secured from injury in chip boxes ;
to these should be affixed labels, stating the place they
were found in, and any other circumstances. Those
armed with long and tender spines had better be enveloped in cotton or tow, until their points are completely
covered : the rest may be wrapped in cotton, paper, or
other soft substance, and closely packed ; taking care
to put the largest and heaviest at the bottom, and filling
up the interstices with the smaller species. Many of
these latter, also, may be packed, with greater security,
within the large ones ; thus the risk of injury will be
diminished, and much space saved.

CHAP. II.

ON PRESERVING ZOOLOGICAL SUBJECTS.

INSTRUMENTS. — CHEMICAL DRUGS. — PRELIMINARY REMARKS
ON STUFFING QUADRUPEDS.

(37.) THE most extensive collections of natural objects are but of little value, unless their durability is
secured by judicious preparation in the first instance,
and by watchful attention after they are placed in the
museum. In most instances, the careful performance of
the first is necessary to secure the success of the last.
During the two preceding centuries, the art of preserving
animal bodies, otherwise than in spirits, was but little
understood ; and to this cause, more than to any other,
must be attributed the partial or total destruction of
those extensive collections of animals made by sir Hans

Sloane, and the great naturalists who lived about the same period ; collections, whose existence we now only read of — for they have nearly passed away " from the things that be."

(38.) The forms of animal bodies may be preserved, by preventing their substance from undergoing decomposition, by executing correct models, or by drawing their representation upon a flat surface. Commencing with the first of these methods, or the preservation of animals, properly so called, they can only be kept in an entire state by immersion in alcohol : this is the usual method adopted for the marine animals, as fish, mollusks, &c., as well as many annulose families, as spiders, scorpions, &c. The preservation of the skin, however, is that most generally useful ; the internal parts being removed and the space filled up by other substances.

(39.) The *instruments* and other articles, with which the operator should be provided, are as follows : — A small case containing knives or scalpels, and scissors with pointed blades, both of different sizes. Forceps, both pointed, obtuse, smooth, and indented. Two flat pincers, large and small ; a round pincer ; a cutting ditto. A hammer. Files for wire, &c. Brushes of different sizes, for applying the preserving drugs, or for smoothing the fur, feathers, &c. A thin goose-quill entire ; and another of a larger size, cut like a scoop, for removing the brains of small birds. A box of powdered chalk for absorbing blood. An assortment of iron wire of several sizes. Needles and silk, flax, tow, or cotton, and a collection of enamel eyes.

(40.) The *chemical preparations* or compositions, used in anointing the skin, are of various kinds. Numerous recipes for these are to be found in the works of Lettsom, Donovan, Graves, and others ; but experience has shown them to be ineffectual, and they are now no longer used. The merits of the three following compositions have been sufficiently established to warrant their general adoption in preference to all others.

1. *French arsenical soap*, invented by Bécœur. —
Camphor, 5 ounces; powdered arsenic, 2 pounds;
white soap, 2 pounds; salt of tartar, 12 ounces; powdered chalk, 4 ounces. Cut the soap into small slices
as thin as possible; put them into a pot over a gentle
fire with a very little water, stirring it often with a
wooden spoon; when dissolved, add the salts of tartar
and powdered chalk: take it off the fire, add the
arsenic, and stir the whole gently: lastly, put in the
camphor, which must first be pounded in a mortar with
a little spirits of wine. When the whole is properly
mixed together, it will have the consistence of paste. It
may be preserved in tin or earthenware pots well closed,
and cautiously labelled. When wanted for use, it must
be diluted with a little cold water to the consistence of
clear broth: the pot may be covered with a lid of pasteboard, having a hole for the passage of the brush by
which the liquor is applied.*

2. The *arseniated soap*, used by us in South America,
is thus composed: — Arsenic, 1 ounce; white soap,
1 ounce; carbonate of potash, 1 drachm; distilled
water, 6 drachms; camphor, 2 drachms. This mixture
should be kept in small tin boxes: when it is to be
used, moisten a camel's hair pencil with any kind of
spirituous liquor, and with it make a lather from the
soap, which is to be applied to the inner surface of all
parts of the skin, and also to such bones as may not be
removed.† When made up into cakes, this composition
has all the appearance and solidity of common soap.
Hence it is more adapted for travellers, as being in a
less fluid state than the former; and one piece, no larger
than an ordinary cake of Windsor soap, is sufficient to
preserve 500 small birds. Great care, however, must
be taken in using this, as well as all other similar compositions. If the least particle gets between the skin
and the nail, and is not immediately removed, it sepa-
rates both much lower down than their natural limits,

* Taxidermy, p. 16.   † Naturalist's Guide, p. 63.

creates great pain, and renders the fingers very tender. We should therefore recommend the operator to wash his hands and clean his nails immediately after he has finished applying it to his specimens.

3. *Bullock's preservative powder.* — Arsenic and burnt alum, each, 1 pound; tanners' bark, 2 pounds. Mix the whole, and, after reducing it to powder, pass it through a sieve; finally, add half a pound of camphor, and half an ounce of musk: let the whole composition be well mixed, and kept in close tin canisters. This powder is more particularly adapted to fill up incisions made in the naked parts of quadrupeds and the skulls of large birds. It has been strongly recommended to us, but, being perfectly satisfied with our own, we have never tried it.

(41.) The following *glue*, in the *mounting* or *erection* of vertebrated animals, will be found useful for a variety of purposes. Take half a pound of common gum arabic and two ounces of white sugar candy: melt this mixture in a pot of water, and strain it through a linen or a horse-hair sieve. When it has become liquid, put a part of it into a flat preserve pot, add a spoonful of starch or hair-powder, and mix the whole well together with an iron wire which should always remain in the pot for this purpose. This gum never spoils: when it becomes dry, add a little water to it; if wanted for instant use, by placing the pot on warm ashes, or in warm water or sand, the gum will very soon melt. The French artists recommend a paper paste, for which they have given us the following receipt :— " Fill a large coffee-pot with water and unsized paper, such as is used for printing; boil it for two hours, renew the water, and boil it again for the same time : then squeeze the paper and pound it in a mortar until it be reduced to a very fine paste: then dry it; and when there is occasion to use it, take some melted gum arabic, add the powder to it, and a large handful of pounded paper, mix the whole well together, and put it into a flat pot." *

* Taxidermy, p. 17.

(42.) There are *certain general rules* which should
be attended to in the preparation of all vertebrated
animals. In warm climates, the operation should commence with as little delay as possible, allowing sufficient
time for the muscles to relax from that tension they
acquire immediately after death; although this delay is
not at all necessary. In tropical latitudes, the animals
killed in the morning will frequently become tainted
towards the close of the day; whereas, in northern
climates, the process of decomposition will not be apparent for several days: in either case, it may be considerably checked by the application of powdered charcoal, either put into the mouth and near the *anus,* or
sprinkled between the feathers or fur. In many cases,
animals may be sent to a considerable distance, if put
into a box of powdered charcoal, well secured: in this
manner the rare aquatic birds of Shetland and Orkney
might be sent to London with perfect safety; and it is
well known that the *Tetrao Urogallus,* or wood grouse
(now extinct in Britain), is annually sent to us from
the coast of Norway, and even from Russia. This
practice has now become so common, that we saw, this
year, a remarkably fine specimen exposed for sale at *5s.*
only, at the game-shop at the bottom of Holborn Hill.

(43.) Previous to making any incision, let accurate
measurements be taken of the subjects; these may be
introduced in its subsequent description, and will also
prevent the operator from stretching the skin beyond
its due limits. The neck, more particularly, is liable
to be so distended, and therefore the natural length
between the tip of the head and the base of the fore
feet or wings should be carefully preserved. The incision or slit by which the body is removed, should not
be larger than is absolutely necessary for that purpose:
practice alone can give a proper degree of skill in this
respect. Nothing tends more to insure the preservation
of a vertebrated animal, than the effectual removal of
every particle of flesh, muscle, or bone, not absolutely
essential to preserve the symmetry of the outward

form: feathers and furs are destroyed by moths; but numerous other small insects, such as *Acari, Ptini*, &c., take up their residence in the interior of all skins where an undue proportion of the fleshy or bony parts has been suffered to remain, which have not been well anointed with some chemical preparation.

(44.) Commencing with *Quadrupeds*, the operator should begin by opening, cleaning, and filling the mouth with cotton or tow, to prevent any blood or moisture from exuding. All wounds should be treated in the same manner. The animal is then stretched on its back, and the hairs being turned to the right and left, the skin is to be opened in a straight line down the middle of the abdomen, commencing from the arch or hollow of the pubis, and ending with the stomach: the upper part of this slit may be extended to the collar bone; but, as the operator gradually acquires dexterity, he will be able to decrease or shorten it. Care must be taken not to injure the muscles of the belly, by making the first incision too deep, otherwise the intestines will fall out and soil the fur. The operator then proceeds to separate the skin from the flesh, both to the right and left of the belly, placing pads of tow or linen between, and sprinkling powdered chalk on the flesh as he proceeds, by which means the slime and blood upon the surface is absorbed: the anus is next detached from the rectum, the tail cut off interiorly at the last joint, and each thigh separated at its junction with the bones of the pelvis. Hitherto the animal has remained upon its back; but it must now be laid on its side, the posterior part towards the left, and the belly towards the operator. In this position, the thighs, being separated, recede towards the right, and give more facility for skinning the back: this last part is always the easiest. For quadrupeds of a small or middling size, it is sufficient to take the skin in one hand, and the body in the other, and by drawing them in contrary directions, to unskin the body as far as the scapulæ, or rather to the shoulders. The arms or fore feet are next separated,

by being cut off internally at the shoulder joint : the
neck and head are exposed by drawing the skin over
the latter as far as the end of the nose, taking care to
cut the ears as near as possible to the skull, and not to
injure the eyelids, or cut the lips too close.   To separate the carcase from the skin, it now only remains to
cut through to the last joint of the neck, taking all the
muscles, and leaving the bones of the head perfectly
clean and naked.   To remove the brain, the occipital
hole must be enlarged by a strong sharp knife, and the
contents of the skull extracted by means of a wooden
scoop.   The whole being well anointed with the preservative soap, the head is put back into the skin.   The
fore legs are then to be cleaned, by drawing them towards the operator, and pushing the skin the contrary
way; the ligaments uniting the bones must be preserved, but the whole of the flesh removed ; the legs
should then be returned into the skin.   The hind legs
are to be treated in the same way; skinning them as far
as the claws, preserving the ligaments, and removing the
flesh.   It now only remains to skin the tail, which is
generally the most difficult part of the operation, and
various modes must be adopted according to the nature
of its form.   With monkeys, and other long tailed
quadrupeds, the first two or three joints must be laid
bare, and strongly tied with a cord, the other end of
which is fastened to a hook or other immoveable body :
the bony joints of the tail are then separated from the
skin by means of a cleft stick, which is passed between
the cord and the skin ; the bare joints passing within
the cleft ; and by drawing the skin towards the extremity, the tail comes out of its sheath.   The skin is then
carefully extended on the table, with the legs stretched
out ; and when all the remaining muscles and particles
of flesh are removed, and the inside well anointed, it is
ready for stuffing.

(45.) For *mounting* quadrupeds,— that is, for giving
them their natural form and attitude,— the following process, practised in the French Museum, is recommended.

Supposing the subject to be of the size of a fox, let some iron wire be selected of such a thickness that four pieces, introduced into the legs, will be sufficient to support the animal. A thinner piece, of about two feet long, is next taken, and bent at nearly one third of its length into an oval shape somewhat smaller than the hand: the two ends are then twisted together, leaving one end a little shorter than the other: then, measuring the iron by the skinned tail, it must be cut the same length, independent of the oval. The wire is next wrapt in flax, — taking it by the point, and turning it round between the fingers; constantly increasing the flax towards the oval. Rub the whole with a little flour-paste, to preserve the shape; and it then should be of the length and circumference of the skinned tail : it must afterwards be left to dry. A little of the preservative may be introduced into the tail with a small brush; and the towed wire above described (and which may be called the tail-bearer), may be treated in the same way, and put into the skin of the tail. The oval end, which is now placed within the body, serves to fix the tail to the iron which represents the back-bone. Five pieces of iron wire, of the diameter of a straw, are next to be selected : one of these must be a foot longer than the body of the animal ; the four others should be of the same length as the legs, which they are intended to support ; the points of all must be sharpened at one extremity in a triangular form, in order to penetrate the more easily. At the unpointed end of the longest of these wires, a ring must be formed, large enough to pass the little finger through ; bending the wire back on itself a turn and a half with a pair of round pincers. A similar ring on the same wire must be formed (by one entire turn) in that part which will come between the animal's shoulders : the rest of this wire must be perfectly straight, and pointed triangularly at the end, as before described. The irons being thus prepared, and the skin of the animal extended on the table, the end of the nose is taken with the left hand; and thrust-

D

ing it again into the skin, the bony head is received
with the right hand, which has been introduced into the
neck.   Having anointed it with the preservative, all the
cavities and hollows, where flesh or muscle had existed,
must be filled up with chopped flax : one end of the
long piece of wire is then introduced into the middle of
the skull, and the head is then restored to its proper
place : the inner surface of the skin of the neck, being
likewise anointed, is to be stuffed with chopped flax,
always paying particular attention in referring to and
preserving the natural dimensions,—as all fresh skins
easily dilate, but can never be effectually contracted
when once dried.   It has been before observed, that the
first ring of the wire, which passes into the head, must
be in the direction of the shoulders ; the second ought,
in like manner, to correspond with the pelvis, or a little
towards the posterior part.   One of the foot-wires must
next be taken and passed behind the bone of the front
leg : the point which comes out at the sole, should be
under the highest ball of the foot.   This done, the bones
of the leg are drawn up within the skin of the body, to
tie the iron wire to the bone of the arm and fore-arm
with packthread.   The parts are to be anointed and twisted
with chopped flax, observing to make the thickness proportionate to the flesh that has been removed.   The
operator now proceeds to fix the fore legs, which is done
by passing their wires in the little ring of the middle
or back-wire, and twisting the two ends strongly together by the help of flat pincers.   For an animal of
the size of a fox, the pieces left to twist should be from
5 to 6 inches long.   This done, they are to be bound
on the under side against the back-wire, and fastened
together with packthread.   The two legs are then replaced, and bent according to the attitude intended to
be given to the animal.   The skin of the shoulders and
belly are next anointed and stuffed; care being taken to
put a sufficient layer of flax under the back-wire.   The
operator must now begin to sew the anterior part of the
opening ; care being taken to preserve the external ap-

pearance of scapulæ, and more particularly that thickness which appears beyond, at the junction of the
shoulder and bone of the fore foot. The hind legs are
next to be commenced upon. The wires for these ought,
in general, to be longer than those for the fore legs; they
are to be inserted into the paw, and loosely fastened to
the thigh and leg bone: the flax, by which the natural
form is to be restored, is to be applied as before directed;
and if the whole is bound round with thread, it will
prevent it slipping up when the leg is returned to its
natural position within the skin. The hind legs are then
fixed, by passing the extremities of their wires in the
second ring of the back or central wire, which ring
should be situated at the pelvis: the two ends are then
bent, and twisted in opposite directions round the ring.
To give additional strength to this part, a piece of packthread may be passed several times round these three
wires, and strongly tied. The tail-wire is then to be
placed in the manner already described. The internal
iron work is now done, and it only remains to anoint
all the interior parts of the skin still exposed, and to replace the body of the animal by chopped flax or other
soft substance; laying it conveniently under the wires;
carefully preserving the natural circumference, and imitating, as much as possible, the superficial irregularities
caused by the muscles in the living subject. Lastly,
being provided with a proper-sized triangular-pointed
needle (called glovers' needles) and strong silk, proceed
to sew up the longitudinal incision down the belly;
passing the needle from the inner surface, and taking
care to divide the hairs to prevent their being drawn in
with the edge of the skin.

(46.) The actual *erection* of the quadruped, thus prepared, is the next process. When the skin is sewed up,
the subject is to be turned in all directions, and kneaded
or pressed by the hand in every part, in order to model
it into a more correct shape, and restore as much as possible the appearance of the various muscles. A board
is next to be taken, in which four holes (of the same

circumference as the foot wires) are to be drilled, at distances suitable to the attitude intended to be given to the specimen. The animal is then placed upon the board, the wires of the feet inserted through the holes, and drawn by the pincers so close to the board, that the soles of the feet rest firmly upon the plank : the ends of these wires will then project on the other side of the board; they must consequently be well bent and clenched with short nails. The board now lying flat upon the table, and the specimen erect, the operator proceeds to give the proper attitude to the head ; imitating the appearance of muscles by stuffing in additional cotton at the orifices of the eyes, mouth, ears, nose, and anus : if any part appears hollow, a piece of strong wire, hooked at one end, will draw forward the flax inside, and remedy this defect. The artificial eyes are to be put in while the eyelids are still fresh. This operation should be done with care and neatness, the curve of the eyelids well preserved, and sufficiently drawn over the eye to give the natural fulness in that part not exposed. The artificial eye should, of course, perfectly correspond in size and colour with that of the living animal. A proper degree of plumpness is to be given the lips by pads of cotton placed inside and secured by pins : the nostrils are likewise to be well filled, and naturally distended by cotton closely pressed, and the flesh completely saturated internally with the preservative. If the ears are to be erect, a connecting thread is passed through the base of each, and tightened until they are sufficiently near to each other. If the ears are large, a piece of pasteboard of the same form may be placed within, and fastened round the edges with small pins; or a thin piece of cork, if at hand, will answer this purpose better.

(47.) A quadruped may be thus prepared in four or five hours, and will remain uninjured for a great number of years. As a measure of precaution, lest the preservative liquor may not have sufficiently penetrated the naked parts, such as the ears, nose, lips, and paws, they may be anointed with a brush dipped in spirits of turpentine. That this liquor may not injure the hair, the latter must be wiped afterwards with cotton. The operation is to be repeated seven or eight times, at intervals of some days. When the animal is quite dry, the wire which passes from beyond the head is to be cut with pincers. It may either remain on the original board, or be transferred to another, perforated in the same manner, and the wires which support it securely riveted.

(48.) Several variations from the above process are made to suit particular tribes of quadrupeds. Bats may be prepared without wire, and a flying position given to them, by extending the wings on a piece of soft wood, to which they can be fastened with pins, and remain until dry. In bears and other large animals, the back wire gives place to a piece of wood, as being stronger. The tails of such quadrupeds as are very thick (as the beaver and African sheep) must be cut underneath, and the flesh removed. Deer and other horned animals are skinned in the usual manner, so far as the neck, which is cut off as near as possible to the head : another opening is then made, beginning at the chin, and continuing it down the neck until it is 8 or 10 inches long : by this opening the remainder of the neck is removed, the tongue taken out, and the occipital hole enlarged. The lips are then cut as near as possible to the jaw bones : the skin is then entirely separated from the head, except at the muzzle, where it is left to adhere. The head, being well cleaned, is then anointed, filled with chopped flax, and the skin carefully replaced : the opening is sewn up with very small stitches, that the hair may cover and conceal the seam. Quadrupeds of the largest size are generally mounted upon a wooden model.

(49.) When the skins of quadrupeds are not intended to be mounted at the time of preserving them, the tediousness and difficulty of the process is much lessened. In such cases, it is only necessary to remove the flesh and muscles of the head and legs; anointing them and their bones with the preservative composition, which must

also be well applied to the skin. The specimen may
then be stuffed and sewed up : the extremities, however,
are to be well saturated with spirits of turpentine, and
the application repeated in three or four days. The
skulls, when wanted for anatomical purposes, may frequently be entirely removed from the skin, and their
place supplied by a wooden or cork model.

(50.) We now come to BIRDS, the most beautiful of all
vertebrated animals.—The first precaution to be taken is
to cleanse the mouth of blood and mucus, by means of
cotton affixed to a wire, bent at one end. The specimen
is then to be laid on the table or dissecting board, the
head being turned towards the left of the operator. The
feathers of the belly are divided, in opposite directions,
in a straight line ; and the down, which will be then exposed, is removed by a pair of small forceps : an incision
is then made in the skin, from the commencement of the
sternum or breast-bone until beyond the middle of the
belly : the skin is then raised on one side by the forceps,
and separated from the muscles by a knife; continuing
to do this as near as possible to the wings. A little flour
or powdered chalk sprinkled on the skin and flesh will
prevent the feathers from sticking to them. The thighs
are then forced up or pushed out within the skin, and
are cut between the femur and tibia, in such a manner
that the former remains to be pushed back into the skin.
By the help of the knife and the fingers, the skin is detached as far as the rump, which is cut off from the
body, as it remains to support the tail feathers. The
uncovered carcase is then taken with the left hand, and
the operator continues to separate the skin from the two
sides : the little tendons which are found near the wings
are cut with scissors : the wings are separated from the
trunk at the junction of the humerus, and then restored
to their proper place. The operator continues to skin
the neck, thrusting the head from within in the same
manner as is practised with quadrupeds. In uncover-
ing it, great care is taken not to enlarge the opening
of the ears, or to injure the eyelids in removing the eyes,

which are to be taken out with the closed points of the scissors, and the cavities filled with cotton. The neck is then separated, the tongue taken out, and the flesh and muscles between the two branches of the inferior mandible removed. The brain is got at by enlarging the occipital hole, and extracted by a scoop cut like a toothpick, which can be made either from a quill or reed: the interior of the skull is well cleaned, anointed with the preservative, and filled either with tow or flax. During all these operations, the skin should be sprinkled either with powdered chalk or fine dry sand, to prevent the feathers from adhering or becoming dirty. The wings are now taken out, and cut off at the second joint — that is, the joint next the shoulder, — the muscles removed, and the wings restored to their natural position after the bones have been anointed. In like manner is to be removed the flesh and muscles of the thighs, preserving the bones of the legs, and then replacing them.

(51.) If the size of the bird renders it necessary, all the muscles which adhere to the skin, as well as the fat, must be carefully taken away; and any holes, formed by the shot, should be sown up. Lastly, a piece of thread is fixed to the first joint of each wing, by which they are drawn together to the distance they occupy when the bird is in flesh. This precaution, which does not appear of any great importance, infinitely abridges the trouble of the subsequent operation ; for when the bird is mounted, the wings place themselves, provided they are properly tied within.

(52.) The process of *mounting* birds, prepared in the above manner, is the next step. The head is to be replaced within the skin (previously anointed with the preservative), by holding and gently pulling the thread which ties the beak with the left hand, and assisting its passage into the skin with the fore-finger of the right hand. The feathers can be arranged with a pin or needle fixed within a handle. The bird is then laid on the dissecting board, the head to the left, having properly arranged the wings and legs. To keep it in this position, a leaden weight is placed upon the tail. The feathers on the edges of the longitudinal incision are then raised, for the purpose of anointing the interior skin of the neck, in which " the preservative is introduced alternately with the flax, without stuffing it too thickly," which is a fault generally seen in mounted birds. The operator continues to anoint the back as far as the rump, stuffing it nearly one third of its thickness, that the iron wires may be placed on a thick layer of flax. Four wires are then to be prepared, the proportions of which are as follows: — The first, or back wire, should be longer than the body of the bird; at about a quarter of its length, it is twisted into a small ring by the pincers; the other extremity is pointed. Two others are to be somewhat longer than the legs, which they are intended to support: these we shall call the leg wires. The fourth, or tail wire, is to be formed into an oval by twisting the ends, two or three times, in such a manner that, after being twisted, these two ends form a fork, and the oval is nearly the third of the length of the bird's body : the two teeth of the fork must be pointed with a file, and near enough together to enter the rump; their ends will be hid under the great feathers of the tail, and the oval in the body of the bird. These wires being correctly fashioned, they are to be thus applied: — The back wire, being oiled, is to be introduced " across the skull, passing it into the neck in the middle of the flax with which it is stuffed ; so that, having crossed the skull, the ring of the wire is placed a little towards the anterior part, and can receive the extremities of each of the wires which have passed through the thighs and claws, after having been also pointed." *   These leg wires are introduced, by making a passage through the " claw and bone of the thigh" by an awl of the same circumference as the wire. The wire is to be passed in a straight direction over the knee, and, being shoved out within the skin, is to be brought in the little ring of the back wire ; the other leg wire is applied in the same manner; and both

* Taxidermy, p. 59.

these and the end of the back wire beyond the ring are
to be twisted together with a flat pincer, and lowered
towards the tail. In large birds, it will be necessary to
fasten the tail wire to the others, but in small specimens
it may remain free. The wires being thus adjusted, and
resting on a layer of flax, the skin is to be well anointed
in all parts, and filled to its natural dimensions. The
skin must then be sowed up with a triangularly pointed
or a common needle and strong silk; passing the stitches
from the interior surface, so that the needle comes out
in a direction towards the operator.   If the orbits of the
eyes are not sufficiently plump, a little more cotton can
be introduced under the eyelids, moistened with gum,
by which the artificial eyes will be more firmly fixed.
In this latter operation, much care and delicacy are required : the eye should have a moderately plump appearance; and the eyelids be well rounded, and drawn over
the glass.

   (53.) The French artists mount their birds in the following manner : — " In the middle of a piece of square
wood, we fix an upright, crossed by another piece forming a crutch: we pierce the latter with two holes at the
distance which exists between the feet of the bird ; passing into them the two ends of wire which come out under
the feet, and which have been left long enough to turn
them on this cross stick, to steady the bird." The bird
" being on its wooden support, we must press our two
thumbs on the legs or tarsi, to incline the bird backwards;
then bend the tibia to bring the body forward: before
this operation, the tibia and tarsus were in a straight
line ; they now form a natural angle. When it is well
placed, we bend or turn the head according to the attitude
we wish to give the bird ; and afterwards arrange the
wings. It only remains to smooth the feathers into their
natural position ; and, to keep them in place, we encircle
the bird with small fillets of gauze or muslin fastened
with a pin.   When the bird is quite dry, we take away
the fillets, cut the wire of the head as close as possible
to the skull, place it on a new foot of turned wood pro-

portioned to its size, write the names of the genus and
the species on a ticket of white card, and fix it on the
upright of the foot with a little gum."* The frame-
work above described, is stated to be the most simple
and best adapted for small birds. We will mention
another, which answers for the smallest as well as the
largest birds, and which we adopt in preference. It is,
like the preceding, composed of five pieces. The first,
or centre, ought to be nearly twice the length of the bird:
we bend it at a third of its length in the form of an oval,
twist it two turns, then pass the shortest end into the
oval, and raise it against the longer end, so as to form a
ring at the end, or beyond the oval, big enough to receive the two wires from the claws ; we twist it a second
time, uniting it strongly to the long end, which is straight
and painted; then rubbing it with oil, we enter it into the
neck, already stuffed with flax: the oval of the iron
ought to be in the middle of the body. The wires of
the claws must, like the others, be straight and pointed:
we also enter them through the soles of the feet. When
the point is in, we curve it at the other end, to be the better
able to work it up with the hand; and when the point
appears within, we draw it up with the flat pincers, after
straightening the other end. To fix the irons of the
claws to the middle branch, we pass the two inner ends
into the little ring above the oval ; we twist them together, and curve them within; we then fasten them
with a thread or packthread to each side of the oval."
The tail-wire is exactly similar to that which has been
previously described, and is fixed in the same manner ;
" thrusting the fork into the rump, and either leaving
the oval free, or tied under that of the middle wire.
This machinery, although different to the other, is always
introduced after the neck and back are stuffed. † M.
Maugé had a third method of constructing the interior
frame-work for small birds. He selected two wires (proportionate to the size of the bird), one of which was a
little longer than the other: he pointed both ends of the

* Taxidermy, p. 61.   † Ibid. p. 63.

longest piece, and one only of the shortest. He held
one end of each wire under the thumb and forefinger of
the left hand, and at about the distance of two thirds of
an inch he twisted the other parts five or six times with
the same fingers of the right hand ; after which he left
a space untwisted large enough for a finger to pass
through : he continued to twist it four or five turns,
leaving a second interval untwisted for the passage of
the two wires of the claws : he then gave the form of a
triangle to the first space. We conceive that the smaller
opening or second distance ought to be one turn above
the triangle. The two leg wires are formed in the common way: to fix the back wire when the head and neck
were stuffed, he introduced the long end through the
neck and skull; the fork at the other extremity passed
across the rump to support the tail ; one of the leg wires
being then passed up, he brought the end through the
little hole above the triangle; he bent it along the opposite part, and united the two parts by tying them with
thread: both legs were done in this way. For large
birds, M. Maugé formed the back wire on the oval principle.

(54.) The method of preparing bird skins adopted
by us, and elsewhere recommended *, is as follows : —
" The wings, neck, and joints of the bird should be
rendered lax by moving them backwards and forwards;
the throat is then cleaned with a little cotton, a small
quantity put in the mouth, and a little forced into the
nostrils ; by this means the feathers will be preserved
from being soiled by blood or mucus in the subsequent
process. The bird is then laid on its back, the feathers
divided in a straight line from the breast to the belly,
and the body cut in that direction just deep enough to
divide the outer skin without injuring that which con-
fines the bowels. Proceed to shove the skin gently
from the flesh with the finger, until there is enough to
take hold by: it may then be slightly raised, while,

* Naturalist's Guide, p. 20.

with the other hand, the operator goes on to separate
the skin from the flesh, by pressing the latter with a
blunt quill: this is to be continued until the thighs
begin to appear,— sprinkling the exposed parts of the
body with powdered chalk, to prevent the feathers from
adhering: a little cotton placed between the skin and
body will likewise answer the same purpose : if the
subject bleeds much, a greater quantity of chalk will absorb all the blood, and may then be shaken off in flakes.
Detach the skin sufficiently round the thigh to admit of
the joint of the knee being cut underneath, leaving the
second joint of the leg attached to the skin, and the
thigh joint to the body.   The other side of the bird is
to be treated in the same way.   When this is done, the
skin is carefully detached from the lower part of the
back, the rump is then cut off just above where the
roots of the tail feathers are felt, the skin sprinkled
with chalk, and the naked parts of the carcase covered
with cotton.

(55.) The bird is now to be laid on its side, and the
skin removed from the breast until the shoulder joint is
exposed: at this part, the wings are to be separated in the
same manner as the legs.   The skin is now gently drawn
over the neck, previously passing a piece of strong silk
through the nostrils for closing the bill, leaving the ends
of a convenient length.   When the skull begins to be exposed, the ears will appear like a little hollow on each
side ; they are cut through by passing the point of the
knife beneath, sufficiently deep to scoop out the skin.
The eyes will next appear, covered with a white filmy
skin: this must be first cut through ; and the eyes are then
taken out by cutting the skin all round the socket, and
gently forcing out the eyeball with a blunt stick : this
operation should be done with great care; for if the eye
be perforated, the humours run out and generally soil
the plumage of the head.   The carcase is now completely
separated from the skin, by cutting off the neck where it
joins the skull, the hole at the back of which is enlarged,
the brain extracted by a toothpick quill, and the in-

terior effectually cleaned by a little cotton worked about by a blunt stick.   With a small knife remove all the flesh from the temples, roof of the mouth, and jaws: the ligaments must, however, be left; and the tongue (unless large and fleshy, like those of parrots) on no account is to be removed.   Anoint all the parts with the preservative, and fill the skull with fine tow or flax.

(56.) The head is now to be restored to its natural position.   At all times this requires delicacy, and generally much patience.   The operator begins by taking the skull in both hands, and with the two thumbs gently and gradually forcing the skin back again over the head, all round the circumference.   It frequently happens that this becomes most difficult when it is just about to be accomplished: the skin is then prevented from passing over the protuberance of the lower jaw, which, in such cases, may be pared down sufficiently to admit the skin to slide over.   The head and neck being returned to their natural position, take a blunt needle (fixed in a handle), and smooth the feathers; first lifting them up in a contrary direction, in order to make their roots pliable.   The flesh and muscles of the wings are next to be removed: the shoulder bone is taken in one hand and gradually drawn out, the skin being in the mean time separated by the pressure of the thumb nail of the other hand.   When by this means the joint is exposed, the flesh and muscles are cut away, and the bones anointed with the preservative.   In large birds, it will either be necessary to expose the second joint, or to make an incision along the bone, through which the fleshy parts can be extracted.   Wash the skin and bones, but put no cotton within, as it will never have a natural appearance.   The legs are next drawn out in the same way as far as the knee joint; and, after being cleaned and anointed, are twisted round with tow or flax to imitate their natural thickness: the fat and flesh is then scraped from the inside of the rump, and this part well anointed: lastly, whatever portions of flesh may still adhere to the skin, are to be taken away, and the inside surface completely anointed in every part with the preservative.

(57.) The bird being now in a fit state for *stuffing*, prepare a lengthened piece of tow or cotton, rather longer than the neck, and rolled between the palms of the hand to give it the same natural shape and thick-ness. One end of this tow or cotton is firmly twisted round a stout iron wire, and passed by the mouth into the neck : the wire is taken out at the other end, but the stuffing remains. The mouth is then closed and se-cured by twisting silk round the bill: the other extremity of the false neck is then shortened to the natural length ; and the operator begins to stuff and sew the skin in the usual manner—that is, using a triangularly pointed needle, which is to be forced through the skin from the inside. The bones of the wings are brought into their natural position, and those of the legs so far drawn up that the bend of the knees are in a horizontal line with the vent. The feet and bill, as well as the wattles or other naked parts in certain birds, are anointed with arsenical soap, mixed with spirits of wine ; but some use spirits of tur-pentine. The feathers being smoothed, by turning them upwards and then adjusting them, place the wings flat on the sides of the body, and insert the head into a cap of paper, of such a size that the paper reaches as far as the breast and fits all around. The bird should then be gently pushed upwards: this will bring the neck near to the breast, and make all the feathers lay compact and smooth. Afterwards the specimen may be laid straight upon a board, spread with cotton or tow, where it can remain until dry, and the cap may be taken off. Thin necked birds, as parrots, ducks, woodpeckers, and others having a large head, must have another incision made for the purpose of cleaning out the skull. This second incision need not, in general, be more than three inches in length, and may either be made at the back of the head or immediately under the chin. In large parrots, macaws, and cockatoos, it will frequently be necessary to open the bird from the chin to the vent, or down the whole of the length, because the skin of the neck can by no contrivance be passed over the head.

(58.) Bird skins, thus prepared, may be relaxed, and at any future time be mounted by the following process : — Let a deal box be made of a convenient size (say about two feet square, and proportionably deep), the top of which lifts on and off without any hinges or fastenings : the sides of this box, and the top and bottom, are to be covered with a coating of plaster of Paris between two and three inches thick.   When it is wished to relax any skins, pour into the box, overnight, a sufficient quantity of water to saturate the plaster ; in the morning, any water that is left can be poured off, the bottom dried, and the birds placed within : the lid of the box, being furnished with a groove, will shut close, and the wooden sides will prevent any evaporation going on externally. The box may be placed in a damp part of the house ; and in twenty-four hours, more or less, the skins within will be found perfectly soft and pliant.   This ingenious contrivance, which has never been made public, was communicated to the author by Mr. Bullock, who constantly made use of it for years.   As the moisture would not, in all cases, be sufficient to render the bill and feet perfectly pliable, these parts may be twisted round with wet rags or tow.   If the skins are not sufficiently lax, the seam of the bodies should be unpicked, the inside stuffing taken out with a crooked wire, and the skins again placed in the box for another day.   The French method, for accomplishing the same object, is to fill the skin (after it has been emptied of its former stuffing) with wet pieces of tow or linen ; wrapping the same round the head, bill, and feet, and enveloping all the specimen with a damp cloth until it is perfectly lax.   We have tried this plan, and found it but a rude and inadequate substitute for the last.   The skin is never equally relaxed in all its parts, while the damp of the cloth injures and soils the feathers. If, however, the birds are of a large size, and of any of the marine families, the plan is less objectionable.   In either case, the object being effected, the skin is to be

stuffed and mounted in the same way as those of fresh
or recent birds.

(59.) The process of mounting birds, feather by feather, is resorted to, when valuable skins are too much
injured or decayed to be erected in the usual manner.
For this purpose, a piece of pliable iron wire is taken
of a length proportionate to the bird ; an oval is made
at one end ; and a quantity of flax rolled over the
wire sufficient to imitate the size and form of the
natural body, — occasionally anointing the flax, during
the process, with flour paste. One end of this wire is to
project sufficiently to form the neck, and to be enveloped
for that purpose with flax. This false body is then to be
modelled as near as possible to what may appear to have
been the natural size: the flax, being damp with the
paste, renders this part of the operation by no means
difficult: the model is then dried by the fire, or in the
sun. Meanwhile the head, wings, tail, and legs, are
softened by the usual methods ; the eyes are fixed ; the
wings and tail restored to their natural form by leaden
plates; and the wires passed through the legs, leaving the
ends long: these several members are then fitted and
adjusted upon the false body : if the model is too large
in any particular part, it is reduced by sharp scissors or
any other suitable instrument ; if, on the contrary, it
appears too small, it is increased by gummed cotton or
flax. The ends of the leg-wires, which project from the
thighs, are then introduced through the false body, so
effectually as to come through at the other side (that
is, at the back) sufficiently to admit of their ends
being twisted : what remains may then be cut off: the
other ends of these wires, which come out at the soles
of the feet, are then passed through a horizontal stand,
upon which the bird is to rest, and secured in the usual
way. To render the subject more firm for the subsequent process, an additional wire is passed through the
stand, between the feet, and thrust into the false body:
this temporary support is afterwards removed: a hole is
then made at the rump, large enough to admit the base

of the tail, which is well fastened all round with gummed cotton.

(60.) When it is necessary to paste on the feathers one after the other, the. cross bar or stand which is clasped by the claws of the bird should be separated from the upright part, " and stuck into that of a machine which it is difficult to describe otherwise than by comparing it when it is least complicated to a wooden candlestick, the foot of which is very heavy, and the stem very strong. Several holes should be bored in this stem; one of which, five or six inches from the base, should completely perforate it horizontally, having a diameter of one third of an inch ; the others, of the same size, should be placed obliquely all round the stem, either above or below the first. The ends of the cross bar upon which the bird is fixed should be five inches longer than the bird on each side, and smaller than in the centre, to be able to thrust them firmly into the holes of this candlestick, that the specimen may not be shaken while the feathers are putting on. Being fixed to the new foot, the belly of the bird must be upwards : an amalgam for pasting the feathers is then made of a little melted gum arabic, an equal quantity of the preservative, and a little hair powder ; we put this paste on the belly, and begin by fixing the feathers which cover the tail underneath, then the feathers on the belly, always advancing towards the breast, observing not to lay them on too thick, least there may not be enough to finish it. We must epecially take the precaution of putting the feathers on the places they ought to occupy on the living bird. each on its proper side, because the beards of the left feathers are directed in a contrary way to those of the right; also to observe the shades and dispositions of the colours in the natural bird, and to use them advantageously in the factitious one. We must be careful not to place more than one feather at a time, and to cut the quills of all, to allow the paste to insinuate itself. After having done all this, we give the bird its natural position ; placing the

E

back upwards, we put the gum on the rump, and paste
the feathers which cover it.   Before we paste the dorsal
feathers, we fasten the wings with gummed cotton, and
secure them strongly with pins : we paste on the sca-
pulary feathers in the same manner, and then all those
of the back.   The head will be pierced by the wire of
the neck ; we pull it down to a proper distance, and fix
it by introducing gummed cotton into the skull and
round the neck, which will be made larger than nature,
as it will shrink in drying, and is to receive the shortest
and least downy feathers ; we continue to paste the rest
of the feathers until they mingle with those of the
head." *

(61.) The foregoing process, however tedious and
difficult, is absolutely necessary on many occasions;
particularly for mounting the rarer species of Paradise
birds, and others sent to Europe in the rude state in
which they have been prepared by barbarous nations.
Practice and experience will do a gread deal to lessen
the difficulty.   We advise a sort of apprenticeship, by
getting two birds of the same species ; one of which,
having been mounted by the usual method, will serve
as a model for the form of the body and the disposition
of the feathers.   When the skin of the other has been
taken off and dried, it should be torn in pieces, all the
feathers mixed, and then carefully collected in a small
box. The materials being then prepared, with a correct
model to work by, the operator will proceed with less
fear than if his experiments were first made upon a
valuable skin.

(62.) We now proceed to the preservation of FISH.
The impossibility of preserving the beautiful but eva-
nescent colours of fish, and the unsightly appearance
they generally present, whether in spirits or in a dried
state, prevents these animals from being much attended
to by most collectors. Mr. Bullock, whose name will
be long remembered in this country, in conjunction with

* Taxidermy, p. 81.

the museum which bore his name, has a peculiar method
of preparing fish which has never, we believe, been
made public. His specimens were so perfect, both as
to shape and colour, that they gave the idea of having
just been taken out of the water. A collection formed
by this enterprising naturalist and traveller on the
shores of Mexico, and exhibited in London during 1824,
is still fresh in our recollection, as presenting the perfection of art in this department.

(63.) In preserving fish for the purposes of science,
no method is preferable to that of immersing them in
spirits. The mouth, gills, and fins can then be spread
open ; the rays of the one and the internal parts of the
other can be accurately examined, and even the internal
structure of the body may be investigated. All these
advantages are either partially or totally lost to the naturalist when the specimens have been either stuffed or
dried. It is therefore advisable to preserve all such species as may require subsequent examination in spirits.
Of the vessels to be used for fish and other animals,
glass bottles are perhaps the best, as, whatever precautions are used, a portion of the liquor will evaporate
through the pores of wooden casks : square bottles are
to be preferred, as they arrange close in cases, and no
space is lost. The perfect preservation of the animal
depends upon the quality of the liquor, the manner of
placing them in the bottles, and the method of luting
or closing these bottles. The following instructions
under these heads are taken from the memoir of the
celebrated circumnavigator Peron, whose voyages have
been attended with more advantage to natural science
than any others on record.

(64.) The spirituous liquor to be used must be
from 16 to 22 degrees of the areometer of Baumè ; if it
be stronger, it entirely destroys the colour of the subjects. For quadrupeds it should be of 22 degrees. All
spirituous liquors are equally good, but those which
have least colour are obviously preferable. Before the
fish or animal is put in the liquor, it should be cleaned

from dirt and slime. It is desirable to prevent it from touching the bottom of the bottle, as, if not supported, it will sink down and soon become corrupted. M. Peron therefore proposes to fasten the animal to a flat piece of cork, which holds it suspended in the liquor. Several subjects can thus be placed in the same vessel, either by the side of one another or at different heights; they will float in the liquor without touching, and the slimy particles will become detached and fall to the bottom. M. Peron affirms that thus floating they cannot be injured, although the bottle may be shaken or overturned. But as this method is not very easy, the specimen may be enclosed in a bag of very fine linen, or in a net tied to the cork, to which they will remain suspended. With vertebrated animals it will be advisable to make a small incision in the body, that the liquor may penetrate into the inside. M. Peron advises the use of camphorated spirits, as the camphor augments the preserving quality of the liquor. In some cases, however, this is attended with the disadvantage of making the subjects tough and difficult to dissect. After the animal has been in some days, the bottle must be replenished with liquor, and then firmly closed with a cork-stopper: those made of glass frequently break by the evaporation of the spirits.

(65.) The luting or composition with which the cork is to be covered, and the evaporation prevented, is called by M. Peron lithocolle; it is composed of the following ingredients:—common resin, red ochre well pulverized, yellow wax, and oil of turpentine. The wax and resin is melted, and the ochre added in small portions, stirring it briskly at each addition with a spoon. When the mixture is boiled seven or eight minutes, pour in and mix the oil of turpentine, and continue to boil the whole. Precautions must be taken to prevent the inflammation of these ingredients; but if this should happen, a lid to cover the vessel must be at hand, with which the flame may be immediately extinguished: the vessel should moreover be furnished with a handle, and capable of containing three or four times the quantity of luting that is actually preparing. To ascertain its quality, a small quantity may be put from time to time upon a cold plate, and its degree of tenacity can thus be ascertained. This cement can be prepared at sea, and employed almost immediately. After having corked the bottles, and wiped them well with a dry cloth, the cement is heated to the boiling point, and being well stirred, is applied over the whole surface of the cork by a brush or any other substitute. Sometimes the cement, by penetrating the cork, causes the spirits to evaporate and burst the surface ; this causes small openings to appear, which are stopped by passing a second coat of lithocolle over the first when it is cold. If the bottles are small, the necks may be at once plunged into the cement, provided the grain of the cork is of such a close texture as to prevent any of the spirit from passing through during the operation. The aperture of the bottle may be further secured by covering it again with linen, firmly tied and saturated with liquid pitch. Bottles thus prepared may be turned over in all directions, and exposed to the strongest atmospheric heat without the least evaporation or escape of the spirits.

(66.) The skins of fish may be preserved and dried by different processes. The most simple method, appli cable to the greatest proportion, is that of dividing the fish longitudinally, so as to preserve one side of the skin and fins in an entire state ; from this side the internal bones and flesh are removed. The head is sufficiently reduced in thickness to admit of being laid flat; in other words, accurately divided into two: the dorsal and caudal fins must of course be left entire, as they are too thin to be divided. The skin and bones being well anointed with the preservative, may be either filled with plaster and attached to a board, or be suffered to dry between leaves of blotting paper, and preserved like dried plants.

(67.) Lampreys, eels, and other cylindrical fish may be preserved by skinning them from the head to the tail,

in the same manner as eels are prepared for cooking.
The head, however, must be preserved, its contents
emptied, and the skin filled with fine sand.   The dio-
dons, ostracions, and several other exotic tribes, having
their bodies covered with spines or bony scales, are to be
opened longitudinally under the belly; the interior parts
are then removed, and by being simply stuffed with cotton, the original form is preserved unchanged.

(68.) We are now to speak of REPTILES and AMPHI-
BIANS.   The observations already made on the scientific
advantages derived from preserving fish in spirits, in
preference to every other process, are equally applicable
to all those vertebrated animals usually termed reptiles ;
yet as many from their great size must be exceptions to
this rule, the following methods may be adopted.   In
tortoises, the shell of the back is to be separated from
the breast bones with a strong short knife ; and if the
force of the hand be not sufficient, strike it with a
mallet.   When the turtle is open, take away all the
flesh which adheres to the breast or piece underneath,
and also from all parts of the upper shell ; skin the head,
the fore feet, and the tail, as is done with quadrupeds,
but be careful to leave them adhering to the upper shell ;
pass wires across all the members, washing them slightly
with the arsenical soap, and stuff them with chopped flax
or tow ; then anoint the upper shell with the soap, sew
the parts which require it, and with an awl make four
little holes on the edges of the upper shell and breast
for the purpose of uniting them with thread or twine.
It often happens that the calipash of these animals is
soiled and dirty, in which case it may be cleaned by
rubbing it with a little nitric acid in water ; it may then
be polished by rubbing it with a piece of cloth dipped
in a little oil.*   Lizards of a large size are skinned and
mounted in the same manner as quadrupeds.   Their
colours will fade, but they may be artificially restored,
and the whole covered with two or three coats of varnish,

* Taxidermy, p. 68.

Serpents are skinned by making an opening on the side of the body without crossing or dividing the scales ; the head is preserved entire and anointed, as well as the interior of the skin, with a slight wash of the preservative.   If intended for transportation, they may then be rolled up or pasted on long slips of paper ; if, on the contrary, they are to be immediately mounted, an iron wire is selected of the length of the specimen, round which is twisted unchopped flax, until it attains the natural thickness of the body ; the skin is then extended on a table, and the factitious body sewn up within.   The iron wire being rendered easily pliable, enables the operator to give the specimen any shape he pleases ; when dry the skin may be painted and varnished.

(69.) On the preservation of molluscous and other marine invertebrated animals, little comparatively need be said.   Crabs, lobsters, and other crustacea, when not very large, are transported with least injury in spirits of wine ; or the upper shell may be removed, the fleshy parts taken away, and the specimen suffered to dry in the open air, the preservative having previously been applied.   The flesh in the tail of lobsters is removed by making an incision beneath, and filling it afterwards with cotton.   In some cases, from the great size of the anterior claws, it is necessary to remove the flesh from within ; this can only be done by breaking out a very small piece of the shell, by which the flesh can be extracted, and closing it again by the same piece.   Small species may be dried without any other preparation than remaining two or three hours in fresh water, a general precaution necessary to be observed in drying and preserving all marine substances.

(70.) Other animals of a hard or crustaceous nature, as *Echini, Asteri,* &c. (sea eggs, star fish), are dried in the same way ; the former are emptied by removing the skin which covers the anus, situated on the under part.   These subjects should be suffered to remain in fresh water, changed two or three times, for as many hours, that all the saline particles of their native ele-

ment may be extracted : for want of this precaution the
spines, after a time, fall off, and the essential characters
of the species are destroyed.   Star fish, from their ra-
diate form, can seldom be passed through the necks of
bottles ; they are usually dried on boards or cloths, and
finally anointed with a slight wash of the preservative,
or with spirits of turpentine ; the rays may be made to
dry in any particular direction by the aid of pins.
Corals, sponges, &c. need only be soaked in fresh water
and carefully packed.   The numerous tribes of soft
marine animals, as *Medusæ*, &c., defy all our efforts to
preserve either their colours or form.   Their substance,
indeed, may be retained in spirits; but their parts soon
become so contracted and altered, that no correct ideas
can be gained by a subsequent examination.   The preservation of the mollusca is attended with the same disadvantages ; yet, as our knowledge of the inhabitants of
exotic shells is so very defective, specimens in spirits of
many tribes may throw additional light upon their anatomical construction.   The shells themselves are, perhaps, the most indestructible objects in nature, only
requiring care from violent injury to insure their
durability for centuries.

(71.) ANNULOSE ANIMALS, or insects, require little
or no trouble in preparing ; but the most watchful care
is necessary in their subsequent preservation.   In warm
countries it is necessary to open the bodies of large
beetles, locusts, and dragon flies (*Coleoptera, Gryllus,
Libellula*), by a longitudinal incision, made by fine-
pointed scissors, on the under side, and extending the
whole length of the abdomen ; a small piece of cotton,
fixed on the head of a large pin, will be sufficient to
clear out the contents ; the inner surface of the specimen (if very large) may be slightly washed with the
preservative, and the cavity filled with cotton.   *Libel-
lulæ*, of all sizes, should be treated in this way; and
their bodies, being semitransparent, may be filled with
cotton dyed of the same colour as the insect.   These
variously coloured cottons are used by jewellers, and

give to the specimens a fresh and natural appearance.
Spiders, and other wingless insects with large bodies,
are prepared in the same way.

(72.) *Caterpillars* are preserved in spirits, or by the
following process, practised by Mr. W. Weatherhead:—
The animal is killed in spirits of wine; a small punc-
ture or incision is then made at the tail, by which the
contents of the abdomen are gently pressed out; the
skin is filled with very fine dry sand, and restored to its
natural position. When dry, the sand is carefully
shaken out, and the specimen affixed, by strong gum, to
a piece of card. This seems a more simple plan than
the following, which, nevertheless, is in use among the
French naturalists:—The contents of the abdomen having been pressed out, a slender tube, made from a straw,
is inserted in the orifice, and a pin stuck through the skin
of the caterpillar in a transverse direction, so that it
passes through the straw at the same time, and keeps it
in its place; the subject is then held over a small charcoal fire, but not at its greatest heat, that the posterior
part may attach itself, when drying, to the tube, through
which the operator blows until the caterpillar takes its
proper form, turning it round during this operation
until the skin is dry; the pin and straw tube are then
extracted, or cut close off.

(73.) The *preservation of insects* entirely depends
upon the nature of the boxes that contain them, and the
presence of drugs to deter other insects from attacking
them. In hot climates the ants will find their way to
the store boxes of the collector in less than an hour,
and if the least opening presents itself will commence
their work of devastation. A box of 200 or 300 insects will be destroyed in this way during one night,
and even before some of the specimens are quite dead.
All insect boxes should therefore be air-tight; even
where ants are not to be feared, the cockroaches will destroy all specimens that may be left exposed during the
night. In such situations, insects should be suffered to
dry under the eye of the collector, as the only plan by

which he can insure their preservation. When a sufficient number are ready to fill a store box, it may be
closed (keeping the large and heavy beetles, &c. by themselves), and the lid immediately pasted all round with
paper so as to render it air-tight. If camphor can be
had, a small quantity may be tied up in a piece of
gauze and put in the box. A little alum mixed with
the paste will deter any insects from eating the paper.

(74.) Insects can be *relaxed* at all times, and placed
in their natural attitudes, when their members are sufficiently lax. To accomplish this with specimens that
have long remained unset, a deep basin, filled to within
two inches of its top with wet sand, is provided; the
sand is made perfectly smooth, and then covered by one
or two pieces of blotting paper sufficient to absorb any
water that may remain on the surface; upon this paper
the insects are stuck, and the whole are then covered
over with three or more folds of wet linen. If the
basin is then put into a damp situation, most of the insects will be relaxed in forty-eight hours; but several
will require a longer time before all their parts are perfectly pliable. Many persons merely fix the insects
upon a piece of cork placed in a pan of water covered
over, but this on many accounts is objectionable. Small
Coleoptera are easily relaxed by immersion in hot water,
but those furnished with hair should not be subjected
to this method.

(75.) Insects are *set* or displayed for the cabinet, in
different ways, according to their families. *Coleoptera*,
or beetles, are put generally in a walking position by the
aid of pins and card braces. One specimen of each
genus is usually set with its wings displayed, as if in
the act of flying. And those of a minute or very small
size are fixed (with their legs extended) on slips of
card with gum water. In setting *Coleoptera*, and indeed
all other insects, the antennæ as well as the palpi
(where practicable) should be fully displayed, as being
essential to the scientific examination of the specimen.
Lepidopterous insects are generally set with their wings

expanded, as if in the act of flying. This is done by
means of card braces of different shapes and lengths, by
which the wings are retained in any particular position
until the joints become rigid. The wings are moved
about by a straight or curved pin, fixed into a handle,
and applied to the under surface of the wings, by which
means that exposed to the spectator is neither perforated
nor rubbed. Several species of *Lepidoptera* are liable
to be injured in their colours by the oily matter of their
bodies spreading over the wings. To prevent this in
the first instance, and to remedy it in a more advanced
period, Samouelle recommends the collector " to powder
some fine dry chalk on a piece of heated iron, cover the
chalk with a very fine piece of linen cloth, and thereto
apply the under part of the body of the insect; the
heat of the iron dissolves the grease, while the chalk ab-
sorbs it, and the cloth prevents the chalk from clotting
to the insect. As the under surface of the wings in
lepidopterous insects are frequently very beautiful, and
always different from the upper, it is customary with
entomologists to display one specimen of each species
in this position; that is, with its feet uppermost. One
or two of each genus should also be preserved in the
exact form they assume when the insect is at rest, and
the wings are closed. All other insects are displayed
either walking, or with their wings expanded. No specimens should be removed into the cabinet until they
are perfectly dry.

(76.) The *duration* of preserved insects may be affected very materially by anointing each specimen with
a little spirits of wine, in which corrosive sublimate has
been mixed in the proportion of two drachms and a half
to a pint. This liquor should be applied with a camel-
hair pencil, sufficient only to moisten the under parts of
the head, thorax, and abdomen; as it may have the effect
of relaxing the joints, it had better be applied to old
insects before they are relaxed, and to those which are
fresh, before they are finally deposited in the cabinet.
By proper care, insects may be preserved a great number

of years.   In our collection are several specimens, captured by Bailey, the astronomer, and other naturalists,
who accompanied captain Cook during his last voyage.

(77.) SKELETONS are prepared by two methods,
according to the size of the subject. Those of small
animals are suffered to remain with the ligaments ; but
from the bones of man, and of the larger quadrupeds,
these parts are separated, and the skeleton is united by
artificial means. The first process is the easiest, and that
most generally adopted for animals of the size of a fox.
When the carcass is skinned, as much of the fleshy parts
are taken away as can conveniently be done ; the head is
separated, that the brain may be more readily extracted
by the occipital hole. The bones are then placed in a
vessel of water, to which a little quick lime is added to
accelerate decomposition. After two or three days' maceration (the water having been changed each day), the
skeleton is extended on a table, and all the remaining
flesh is scraped away ; this is repeated until the bones
are completely cleaned, taking every precaution to preserve those ligaments which serve to unite the bones.
As by evaporation the water will diminish, more should
be added, that no part of the skeleton may be exposed,
as it will then acquire a disagreeable blackness. Finally
the bones are scraped quite clean, and washed with lime
water, or a solution of pearl ash (two ounces to a gallon
of water), and then dried in the shade.   Bleaching is
the best method for whitening bones; but with small
animals it tends to decompose the connecting ligaments,
which will otherwise acquire sufficient consistency, when
dry, to support the skeleton.   This object, however, is
better effected, in general, by means of an iron wire.

(78.) To *mount* the skeletons of small animals, the
iron wire last mentioned is passed through the vertebræ,
one extremity being neatly fastened to the head and cer-
vical vertebræ. Other wires are introduced in various
directions, according to the position intended to be given
to the specimen. Should any of the ligaments give way,
and the bones become detached, two holes are made at

the connecting extremities, and they are again united by
one or two twists of brass wire.

(79.) The skeletons of *man*, and of the larger quadrupeds, require a somewhat different treatment, as the
bones must be artificially united; it is not, however,
essential that they should be separated in the first process of removing the flesh more than is necessary for the
convenience of placing them in the vessel.   Maceration
is to be conducted in the same way as for smaller animals,
excepting that more time will be necessary.   Holes should
be bored, about the size of a swan's quill, at the extremities of the large cylindrical bones, to give the water
access to their cavities and a free exit to the medullary
substance.   The vessels should be kept closely covered,
to prevent the admission of impure substances, which
will affect the colour of the bones.   When the putrefaction has destroyed the ligaments, the bones are to be
completely cleaned from all extraneous substances.   They
are then to be soaked for a few days in fresh water, and
lastly in lime water, or a solution of pearlash, as before
directed.   To give them a fine white appearance, many
persons prefer boiling them in the solution of pearl ash;
but perhaps the most effectual method, where it can be
adopted, is to bleach them on the sea-shore within daily
reach of the salt water.

(80.) The mounting of large skeletons is attended
with some difficulty.   Commencing with one of the
extremities, the operator, by the help of a wimble or a
lathe, makes corresponding holes at the apophysis or
round end of the bones, which are then united by wire,
having the ends twisted and sufficiently loose to admit
(in some instances) of a little pliability between the
articulations: this must be done until the whole skeleton is completed.   It is then mounted on a proportionate sized board, and put into attitude by the help of
sufficiently strong wires. The bones of the largest quadrupeds frequently require to be united by a firmer
substance than wire: for this purpose two iron pegs
are used, having a head at one end and a screw at the

other ; each screw has a nut, and to each pair of screws
is provided a narrow plate of iron pierced at each end
for the reception of the screws : these screws are received into corresponding holes made in the bones about
to be united ; they are connected by the iron plates, and
strengthened by the nut or female screw. The skulls of
many quadrupeds, from their size, cannot conveniently
be cleaned unless sawed longitudinally in two.

(81.) Skeletons of small animals can be prepared
through the agency of insects.   Mice, small birds, &c.,
may be put into a proper sized box in which holes are
bored on all sides, and then buried near an ant-hill :
the ants will enter numerously at the holes and eat
away all the fleshy parts, leaving only the bones and
connecting ligaments.   They may be afterwards mace-
rated in clean water for a day or two, to extract the
bloody matter and to cleanse them from any dirt they
may have acquired, then whitened by lime or alum
water, and dried in frames or otherwise as may be most
convenient.   In country places M. Pole sometimes employed wasps for this purpose, placing the subject near
one of their nests, or in an empty sugar cask, where
they resort in great plenty : they perform the dissection
with much greater expedition, and equally as well as
the ants: they have been known to clean the skeleton of
a mouse in two or three hours, when the ants would
require a week. *

(82.) MODELS.—The attempts that have been made
to represent animals by this artificial method have
been very limited, and have proved in most instances
unsuccessful.   The general form may indeed be closely
imitated ; but as the larger classes are covered either
with hair, feathers, or scales, substances which, from
their very nature, defy the utmost ingenuity of man to
imitate, it is more than probable this art will never
make any considerable progress, at least so far as regards the great majority of animals.   We are not aware
of any models of vertebrated animals existing in this

* Pole, Anatomical Instructor, p. 105.

country, executed by European artists. Yet, strange
as it may appear, the Indians of Mexico possess this
art to an extraordinary degree. Among the singular
and well-executed toys of their manufacture brought
to England by Mr. Bullock, was a model in wax, about
six inches high, of a Mexican horse: the proportions
were exquisitely preserved, and the hair imitated with
such singular accuracy that it was only on minute examination that the spectator became convinced it was
not real. Modelling, however, may be successfully applied to represent fossil and recent bones, or other
internal parts of animals : those of a small size may be
executed in wax, while greater accuracy and facility
may be attained by making casts of skulls and such
parts as are more bulky. Modelling in wax might be
advantageously employed to perpetuate the forms and
colours of many tribes of naked marine animals, particularly as they are in general simple, and cannot be
well preserved in spirits. The French have been very
successful in fabricating models of the famous shell po-
pularly called the slipper of Venus (*Carinaria vitrea*,
Lam.). For many years the only perfect specimen
known, was that in the National Museum, then valued
at one hundred guineas ; but of late years it has become
comparatively common.

(83.) A singular method of representing birds,
sometimes practised in England, but more commonly in
Germany, may here be noticed. A correct outline of
the bird, if possible of its natural size, is drawn on
pasteboard : the real feathers are then taken, one by
one, their shafts shortened, and laid on the paper in
such a manner as to give a very tolerable representation
of the subject. Not more than one third of the feathers
can of course be used, and these must be fastened at
their base with strong gum : the legs, bill, and eye
are afterwards painted in oil. In cases where skins
of rare birds are so much damaged as not even to admit
of the operation of mounting feather by feather, this
process will at least secure the fragments from total

loss. The effect would be considerably improved were
the outline to be filled up in the first instance, and rendered somewhat convex by a little plaster of Paris.

(84.) Pictorial Representations of animals, or
zoological drawing, in most cases, is the best and the
most general substitute for the animals themselves, and
is highly conducive to aid the inquiries of the naturalist.
If this is accompanied by a full and accurate description,
without reference to any particular system or method,
the subject can almost always be referred, with little
danger of error, to its proper station in nature.

(85.) It is much to be regretted that, until very lately,
zoological painting has been comparatively neglected.
No principles were laid down by which drawings of
natural objects could be rendered permanently useful or
valuable to science ; the delineation of those parts which
did not immediately meet the eye in the general representation were omitted, while, if the colouring was faithful, and the general outline tolerably correct, both the
artist and the naturalist conceived that every object was
gained. Hence has arisen the numerous mistakes of
authors respecting the identity of a bad figure, the mis-
quotations of compilers, and the tedious complexity of
unravelling synonyms. From this carelessness about
detail, the drawings of many artists, celebrated in their
day, have almost given rise to more error than information ; while, as regards their scientific utility, they have
become of little or no value.

(86.) In *zoological painting*, the first requisite, as regards its application to the purposes of science, is accuracy in the detail. In the higher departments of art,
more particularly in landscape painting, it is only sufficient to imitate general appearances or effects, leaving
the details — such as the form of leaves, the pebbles of
a road, or the tiles on a house — to be filled up by the
imagination. The same principle extends, though in a
less degree, to historical design ; and in both the fancy
of the painter may, to a certain extent, be indulged. But
in proportion as we descend to particulars, our imagin-

ation must be confined, and our accuracy redoubled. If
the painter wishes to make his subject intelligible to a
scientific observer, the smallest tooth of a bat must be
rendered apparent; the exact shape of the scales on the
foot of a bird, and the spurs on the tarsi of an insect,
must be exactly copied. It may be almost said, that no
artist can hope to gain a high degree of perfection in
zoological painting, without being himself a naturalist;
or at least knowing, in a general way, what are those
parts of his subjects which more particularly demand his
attention. These we shall therefore briefly notice, as
they regard the grand divisions of nature.

(87.) In *quadrupeds*, the profile of the head and of
the muzzle, the direction of the ears, the form of the
hoofs, the number and proportion of the toes, and the
size and shape of the claws, are to be particularly attended to in delineating the general figure of the animal.
The following parts should be given separately, and in
detail: — Outlines of the teeth in the upper and under
jaw; of the head in a position different from that in
the general figure; of the toes and claws in a position
best adapted for seeing their peculiar construction.
Where practicable, outlines of the tongue and other internal parts are frequently of great interest.

(88.) In *birds*, the form of the bill, the length and
disposition of the bristles with which in some tribes it
is surrounded, the proportion which the length of the
wings bear to that of the tail, the shape and direction
of the crest of the head, where it exists, and of the
scales on the feet, must be all expressed in the general
figure. The details that should accompany ornithological drawings are many, and require the greatest accuracy. We may instance the fieldfare, to exemplify these
more fully. An outline of the bill in profile, including
the feathers and bristles immediately encircling the
base; a vertical view of the same; the notch near the
tip of the upper mandible; a transverse section of the
upper and lower mandible, by which its relative thickness is seen; the nostrils and their lateral apertures;

F

a leg, including the toes, and claws ; a wing, showing the relative length of the primary quill feathers,
drawn from the inner surface and the termination of
the tail feathers, taken in the same position. When
recent specimens can be procured, the tongue should be
represented laterally and vertically. All these parts
should be drawn either the size of nature, or sufficiently
large to be rendered clear and intelligible. Respecting
the general figure of the bird, it is of little consequence
on what scale it is represented, provided the proportions
are accurately preserved ; neither is it necessary that an
artist, engaged in making a series of drawings, should
attach these details to every species, when any two or
more are found exactly to accord.

(89.) In *fish*, the relative position of the fins ; the
number of the rays composing each, — marking the distinction between such as are simple and hard or *spined*
rays, and such as are jointed and branched or *soft* rays ;
the extent to which they are connected by the mem-
brane ; the form of any filaments or appendages ter-
minating the rays placed on the head, or on other
parts of the body ; the position of the vent ; and the
relative length of the under jaw with the upper, — are
peculiarities necessary to be attended to in designing the
general figure.   The details most desirable are, a front
view of the head, sometimes a vertical outline of the
same part, the gill covers, the teeth, and the tongue.
For the lower vertebrated animals, as the serpents, reptiles, and amphibians, the head and teeth are those
parts most necessary to be represented in detail : the
former, in drawings of serpents, should always be shown
in a vertical direction, and the form of the scales accurately copied. In these, and other oviparous quadrupeds,
many of the generic characters are frequently taken
from the scales, &c. on the under surface of the body ;
outlines of which are, consequently, important.

(90.) In *insects*, the parts to be given in detail vary
according to the different tribes. As a general rule,
however, the various organs composing the mouth, the

antennæ, the palpi, &c., require to be represented as they appear when magnified ; likewise a profile and front view of the head, and of the fore and hind legs. Insects are always drawn of their natural size, and, to be perfect representations, require to be finished in colours. Delineations of the larva and pupa of all insects are highly valuable; and regarding the latter, the greatest care should be taken in expressing the mode of its attachment to other substances, and whether the head is turned upwards or downwards, — as this is an important distinction among the diurnal butterflies.

(91.) In the radiated and inferior invertebrated animals, the variety in form is so infinite, that no particular rules can be laid down. In many tribes, dissection will be necessary to detect their true nature. Calcareous shells, which are the covering of mollusks, are among most beautiful objects, both for the cabinet and the the pencil, that the amateur naturalist can select. In drawing these, the artist should pay great attention to the contour of the mouth, to the delicate striæ that may appear on the surface, to the teeth or tubercles that may exist on the inner or outer lip, to the plaits or granulations on the pillar, and to the termination of the spire. In bivalve shells, it is generally necessary to represent the teeth in two, and sometimes three, points of view; particularly those which rise vertically: to these details may be added magnified representations of the hinge of small and minute species: but in univalves, the upper and under side of which are usually figured, no details are necessary. All marine animals, including the inhabitants of shells, may be drawn with peculiar advantage, if kept alive in a vessel filled with sea water. It is then only that they expand or protrude all their members, and they are seen by the spectator in different points of view.

(92.) Drawings, made on these principles, may be either *shaded* in one colour, as in pencil or Indian ink ; *tinted, i. e.* slightly washed with their natural colours ; or *finished* and worked up to the closest resemblance to

nature which art can produce.   Now, it is evident that
a knowledge of forms will be confined or extensive, in
proportion to the number that may be possessed by the
naturalist, either in his cabinet or his portfolio.   To
possess all animals is impossible : drawings, therefore,
supply the place of the originals; and number, not quality, should regulate his views.   Accurate sketches,
slightly shaded, will frequently give him all the information he wants; and if these are tinted with their natural
colours, it follows that the two first methods, just mentioned, of forming a series of zoological drawings, is
decidedly the best.

(93.) And yet the *perfection* of zoological painting
does not depend on producing an exact imitation of
the object ; for, however desirable this may be, it is,
after all, but a mechanical operation.   The next step is,
to study those forms, actions, and habits peculiar to the
individual in a state of nature, and to select such as are
most beautiful or appropriate, either for the display of
colour, or as strongly characteristic of one or more individual properties.   It is here that genius first enters
into the subject.   Every scientific object may be gained
by copying, with exactness, a stuffed skin, as it is seen
in a museum; but, after all, such a drawing will not represent nature.   The gracefulness and beauty of her
forms must not be studied through the medium of artificial preparations.   To illustrate this, let the figures of
quadrupeds in Buffon be compared with those of Landseer, or the plates of Edwards (*Nat. Hist. of Birds*)
with those of Wilson (*American Ornithology*): the one,
stiff, clumsy, and often distorted ; the other, easy, graceful, and natural.

(94.) To *correctness of outline* should be added a
thorough knowledge of light and shade, or rather those
more delicate principles which produce *effect*.   The
zoological painter is tied down to colour : he must copy
the exact tone of every part of his subject ; and cannot, therefore, avail himself of those diversity of tints
for producing effect, which come within the privilege of

the landscape and historical painter. Perhaps Van Huysum and Rachel Ruisch, in flower painting, excel all others : both are remarkable for the elegance of their grouping, the chasteness of their colouring, and the correctness of their drawing. Van Huysums's notions of effect never extended beyond the light on a grape, and a dew-drop on a leaf: all parts were clearly made out, as if he only painted at noon. Rachel Ruisch, on the contrary, not only studied effect in particulars, but in generals; with all the delicate finishing of her rival, she possessed what he did not, — a depth and a relief, which, while it obscured one part of her pictures, gave a ten-fold richness to the other. We have selected these two painters to illustrate our remarks, as their productions are well known ; but the zoological artist will find that the works of Barraband and Audubon, the two most celebrated ornithological painters, will stand in precisely the same relation.

(95.) We shall now give a list of the most eminent zoological painters and engravers, in our estimation, that have lived since the revival of the arts. The names of the principal publications in which their designs appear, are likewise added. Those whose drawings have not been published, are distinguished by a star (*) ; and this † prefixed to the title of the book, denotes that the name of the artist is not known.

1. *Zoology in general.* — England. MAZIL (Pennant's British Zoology). JAMES SOWERBY (Zoology of New Holland, &c.). J. D. C. SOWERBY (Genera of Shells, Zool. Journal). SYDENHAM EDWARDS (Rees's Cyclop.). J. HOWITT (various). — France. M. C. R. LE SUEUR, now in America, the most eminent painter of general zoology of the present day : his designs, etched by himself, in the Philadelphia Journal of Science, are in every respect inimitable. M. HUET, (Férussac Moll. &c.). BESSA (ditto).

2. *Quadrupeds.* — England. T. LANDSEER (Cuvier's Animal Kingdom, by Griffiths). Major HAMILTON SMITH (ditto) THOMAS DANIELS (African Scenery, &c.). HOWITT. SYD. EDWARDS. BEWICK (Hist. of Quad.). — France. HUET (Coll. de Mammifères). — HOLLAND († Johnson's Quadrupeds).

3. *Birds.* — England. Bewick (Nat. Hist. of British Birds).
Edwards (Rees's Cyclopædia) *Howitt. William Lewin
(British Birds). Selby (British Ornithology). Gould
(various splendid publications). Lear (Psittacidæ). —
France. Barraband (Vaillant, Hist. Nat. des Perroquets,
&c.). Mademoisselle Pauline de Courcelles, scholar of
Barraband (Desmarest, H. N. des Tangaras, &c.).
M. Paul Oudart (Vieillot, Galerie des Oiseaux). Petre
(Vieillot, N. H. Ois. de l'Amer. Sep.). J. Lebrecht
Reinold (Vaillant, Ois. d'Afrique). *Pelletier, an eminent artist, settled in London. France has always excelled
in Ornithological painters.—Germany. †J. L. Frisch (Representation of German and exotic Birds, in German, 2 vols.
folio, 1763). — Holland. Sepp (Birds of the Netherlands,
1770—1809, 4 vols. folio).—America. Wilson (American
Ornithology, 9 vol. 4to.). And last, although perhaps the
first in merit, Audubon.

4. *Fish.* — A. Cooper, R. A. (Major's edition of Walton's
Angler). Donovan (N. H. of British Fishes). †Coral-
lines. †Ellis and Solander's Zoophytes. Naked marine animals. Savigny (Mémoires).

5. *Insects.*—England. Moses Harris (Aurelian. Exposition
of English Insects. Drury's Insects). Lewin (Insects of
Great Britain). Abbot (Insects of Georgia). Curtis
(British Entomology, &c.). — France. Guerin, an admi-
rable artist of all invertebrated animals, especially insects.
†Savigny (Mémoires, 1816). — Switzerland. †Jurine
(Nouvelle Méthode pour classer les Hymenoptères). Sepp
(Insects of the Low Countries, in Dutch; some of the
most correct and beautiful plates ever published). †Voet
(Icones Coleopteorum). †Wolf (Icones Cimicum). The
German entomological works are numerous, and the execu-
tion of the figures are generally good.

6. *Shells.* — The Miss Listers (Lister, Historiæ Conch.).
Laskey (Plates in the Wernerian Transactions). *Lewin.
*Agnew. Burrows (Elements of Conchology). *Miss
Fordyce. Sowerby (Genera of Shells). Crouch (Mawes'
Conch.) M. Huet (Férussac, Moll.). †Drapanaud (H.N.
des Moll.). †Seba (Thesauri). †Poli (Testacia utriusque
Siciliæ). The drawings of the late Miss Fordyce exceed
all others we have ever beheld; except, indeed, those by the
celebrated Dutchman, Wiertz, a name little known in
this country, but who is the *Ruisch* of shell-painters: we
possess a collection of near 200 of his drawings in body colours, each of which is a perfect miniature.

(96.) The most extensive collections of zoological drawings in this country are those of the present earl of Derby, major Hamilton Smith, the late sir Joseph Banks, and, we may add, our own. The first is said to be very numerous; as it contains all those of the late general Davies, and of Sydenham Edwards. The second has been stated to exceed 10,000, chiefly executed by its possessor, who has visited the principal museums both of Europe and America for the purpose of delineating their contents. The third is very curious and valuable; it comprises all the sketches of Foster, Parkinson, and others, made during the voyages of Cook; but the major part, unfortunately, are not sufficiently finished to admit of great scientific utility. A valuable collection of conchological drawings was also formed by our regretted friend, Dr. Goodall, late provost of Eton.

CHAP. III.

ON THE FORMATION AND ARRANGEMENT OF COLLECTIONS.

(97.) WE shall divide this chapter under two heads: — first, as regards those principles most advisable to be adopted in the formation of private collections and public museums; and, secondly, the arrangement and preservation of the objects themselves. Collections of natural objects are, to the naturalist, what a library is to the critic or the scholar; yet with this remarkable difference, that the one draws his knowledge from the works of God, the other from those of man. It would be as vain to attempt to assemble all the books that were ever printed in one kingdom, as to collect all its natural productions. A general knowledge of both is only sufficient, preparatory to the study of any particular portion.

(98.) In the *selection* of specimens there are two principles to go upon. If the object of the collector is to possess the most beautiful examples of a species, either as regards intensity of colour, perfection in its preservation, or in its size, he will find a princely fortune requisite to pursue his plan, at least to any extent. For the objects of science it is sufficient that the specimen is perfect, and that it represents the usual appearance of the species. Yet no scientific naturalist will reject a specimen because it may be slightly injured, seeing it is better to have some acquaintance with one of the forms of nature than none at all; at the same time he will be cautious in drawing hasty conclusions from such imperfect sources; the single valve of a bivalve shell should find a place in his cabinet, until a perfect example can be procured; since he can always obtain a correct notion of the teeth in the deficient valve by an impression of the other in putty or wax. Such specimens need not, however, be mixed with the general collection, but kept in drawers by themselves. The same principle extends to birds, quadrupeds, and all animals, since a part is always better than none.

(99.) When a species puts on different appearances at various stages of its growth, it is desirable to procure specimens sufficient to illustrate these changes. The sexes of birds, insects, and many other animals, frequently vary in a remarkable manner. The colours both of insects and shells are equally inconstant, and their variation requires to be exhibited by a series of specimens. Examples of the same species from different localities should be acquired, as tending to illustrate their geographic distribution, and the changes produced by food, climate, or other local causes.

(100.) *Collections of natural history* are of two kinds: —1. Public or national, founded, like libraries, for the general diffusion of knowledge, and open to the inspection and study of all: to these the name of *museum* is more properly applied. 2. Private collections, formed by individuals either for the gratification of the eye, or the advancement of their own particular studies; these are generally called *collections*, or *cabinets*.

(101.) *National museums* should not only possess types of all the generic forms in the several departments of zoology, but as many of the individual species as possible. The natural productions of the nation, and of its colonies, should more particularly engage the attention of its curators. Every object should be correctly named, according to the best and most recent authorities. A zoological library should be attached to this portion of the institution, that nature may be studied both by books and specimens. The whole should be under the direction of professors of acknowledged eminence in their respective departments, and open without reserve to the inquiries of the naturalist, and the inspection of the public. In this respect the Jardin des Plantes, or the national museum of France, is a model of perfection. It is worthy of a great and enlightened nation.

(102.) During the latter part of the last, and the beginning of the present, centuries, the establishment of national museums have engaged the attention, not only of the different governments in Europe, but even those of America. The most celebrated in the world is that of France : next may be ranked the museums of Berlin, Vienna, Holland, Bavaria, Denmark, and Florence. Science and the arts, so far as public institutions are concerned, have long been dormant in Naples, Spain, and Portugal. Yet, under the auspices of the late emperor of Brazil, a national museum has been founded at Rio de Janeiro, and naturalists engaged to collect the productions of that immense and little-known country. Of late years, zoology has made rapid progress in North America. Universities have been founded in all the provinces; professors of natural history and botany appointed to each ; and public museums are now considered a necessary part of these establishments. Of the public museums of Great Britain, in respect to zoology, little need be said. In the British Museum, there are, it is true, vast numbers of specimens, but the majority are so old and faded, that two thirds might be cast out with much advantage. In shells, we believe, it is very rich, but the whole zoological establishment, when put into comparison with that of France and the great continental nations, is confessedly inferior. The collection of native animals, however, purchased of colonel Montagu, is very extensive, and in good preservation. The Edinburgh College Museum excites the admiration of all who have visited it, for the beauty and perfection of the specimens, and the neat manner of. their arrangement. It is principally composed of the well-known and valuable collection of M. Dufrene, which was offered for sale in this country a few years ago. Its purchase by the Edinburgh College has evinced both judgment and liberality. It was offered to the trustees of the British Museum, but declined. The museum of the Zoological Society is remarkably rich in birds and quadrupeds, but we believe the insects and shells, comparatively, are but few.

(103.) *Local museums* have been formed within these few years, in several of the large and opulent towns of England, by corporate bodies, or private associations. The first of these, we believe, in extent and importance, is that of the Natural History Society of Manchester. It is supported by the annual subscriptions of a great number of members. As they lose no opportunity of applying their ample funds to the acquisition of new objects, this museum is likely to become, in a few years, one of the most important in the kingdom. Attached to the Royal Liverpool Institution is likewise a museum of natural history. The zoological subjects are by no means few, and they have, of late, been considerably augmented: it contains some very fine quadrupeds, presented by Mr. Edmonston, from Demerara: as also several rare birds, from the late museum of Mr. Bullock, and of corals from that of the late Mr. Broderip of Bristol. The museum of the latter city, belonging to the Institution, is stated to be very extensive : it is now under the curatorship of Mr. Stuchbury, whose merits we have had occasion to mention in a former volume.

(104.) There is likewise a museum at Exeter, exclusively devoted to the British fauna, of which it is stated to possess many rare and interesting specimens. The Ashmolean Museum at Oxford deserves to be mentioned as much improved, although quite unworthy of that university. There is a Natural History Society at Worcester, and another at Warwick, while others are springing up in most of the provincial towns of the empire.

(105.) *Local or municipal collections*, formed for the purposes of public exhibition, from their nature and extent, constitute another class of museums. The most celebrated of these at present in existence is said to be that at Philadelphia, commenced by Mr. Peale, and since considerably augmented by his son. Our own country has witnessed the accumulation, and unfortunately the dispersion, of two collections of this description. One was the museum of sir Ashton Lever, so passionately attached to natural history, that he expended two fortunes in its formation, and finally became so embarrassed that he was obliged to part with it. He procured an act of parliament to enable him to dispose of the whole by way of lottery. The late Mr. Parkinson was the possessor of the fortunate ticket. This gentleman immediately erected an appropriate and very elegant building on the Surrey side of Blackfriars Bridge, to which the whole museum was removed from Leicester Square. After remaining for public exhibition some years, it was brought to the hammer. M. Fichtel, an agent from the Imperial Museum at Vienna, came over and purchased nearly all the more valuable objects. The museum formed by Mr. Bullock, and exhibited for several years at the Egyptian Hall, Piccadilly, must be fresh in the recollection of many persons. In the number and perfection of its zoological subjects it far exceeded that of sir Ashton Lever, but like that, it was doomed to share the same fate. The sale was very remarkable ; for besides bringing together nearly all the British naturalists, others were expressly sent from Paris, Holland, Vienna, and Berlin, for the purpose of sharing in its dilapidation.

With such a powerful competition, whatever was rare or
valuable sold at a very high price, while our own naturalists stood by, and had the mortification of seeing
these objects pass into the possession of foreigners.
Some few, the result of the voyages of our illustrious
circumnavigator Cook, were rescued from this general
transportation by the liberality of Dr. Leach; and
several others in the ornithological department, of great
scientific interest, were purchased by the present earl of
Derby, then lord Stanley.

(106.) The museum of the College of Surgeons,
and the Hunterian Museum at Glasgow, owe their origin to the two celebrated men whose name one of them
bears. The former is not only rich in every preparation
and specimen that can illustrate the internal structure
of man and animals, but is, without exception, the most
beautifully planned and the most conveniently arranged
museum perhaps in Europe. What is more, the whole
is catalogued, and every fresh subject is put in its place
almost immediately after it has been received. It is no
wonder, therefore, that donations are sent to it from all
parts of the world, while the unwearied exertions and
the high talents of professor Owen make every one
feel delighted in contributing to its numerous but well
arranged stores of knowledge. It may be as well to
observe, however, that the Museum only contains animals in spirits or skeletons; it being formed more as a
surgical than a zoological collection. The Hunterian
Museum at Glasgow is of a very different description.
It is composed of works of art and nature. The latter,
comparatively, are very few, and by no means interesting
to the naturalist, although well enough to attract the
public eye. The Andersonian Museum in the same
town, as a collection of zoology, is far superior. There
is a very good collection of birds, of which some few of
the rarer species we have described elsewhere. * The
shells and fossils are also numerous and interesting, but
there are very few insects.

* Animals in Menageries, Part iii. p. 281.

(107.) *Private collections* are next to be considered. It is usual to imagine that in the formation of these every naturalist may follow the bent of his own fancy, and such undoubtedly is true. Yet, if he is in the pursuit of science, he will derive a lasting advantage from proceeding upon some one regular plan, adapted to facilitate that line of study he may intend to pursue. No collector, however zealous his endeavours, or however vast his plans may be, can ever hope to obtain a tenth part of the productions of nature, even in one of her departments. A selection is therefore compulsatory; he has, consequently, to choose whether that selection shall be guided by certain rules or by the mere direction of fancy. It may be said that no disadvantage will arise from the naturalist's collecting objects which he has not any intention of studying, but this is a mistake: the versatility of the human mind and its constant desire for change is well known; and by having these objects before him he is frequently tempted from pursuing steadily and exclusively those studies he has chalked out for himself, and which, if he aims at any degree of excellence, require his undivided study. As we think this subject deserving of every attention, particularly from young naturalists, we have here analysed the nature of different private collections, and the system or principles which we think should regulate their formation.

(108.) *Private cabinets* or collections may thus be classed under two heads: — 1. Those intended to illustrate some scientific object; 2. Those formed upon no plan, intended merely for the gratification of the eye. — Scientific collections may be either general, partial, typical, local, or economic. A general collection of all the types and species of animals can never be completed. Yet the attempt, however impracticable, should be persevered in by the directors of all national museums. Such institutions should be to the naturalist what a dictionary is to the scholar—a book of reference, where he may hope to find every word or

every object recorded and explained.   But such a plan is
evidently too vast for private individuals, who usually
content themselves with striving to form general collections of some *one* natural division, as ornithology,
entomology, &c.   Yet even this, from the accession of
new discoveries, is becoming every day more impracticable.   The number of birds either described in books,
or existing in the collections of Europe, probably exceed 6000, while to those of insects there really appears
no end.   Macleay estimates the number already in the
cabinets of Europe to exceed 100,000, and this in all
probability does not comprise more than one fourth of
those actually in existence.   General collections, therefore, of any one kingdom of nature, cannot be recommended, as, independent of the expense of purchasing
and the space they will occupy, the time necessary for
arranging and preserving the specimens will prevent
the naturalist from making any scientific use of his
possessions.   In proof of this it may be mentioned,
that the most extensive collections in modern times
have invariably been formed by those who have not
benefited the science by their writings : the two occupations, in fact, are incompatible with each other. *

(109.) *Partial collections* may be defined as consisting of types, illustrating all the orders, families, or
genera of one kingdom or class ; together with the
species of one or more genera or families, which it may
be the intention of the collector to investigate in detail.
The ornithologist, guided by this rule, would procure

* In conchology, a general collection, or at least the advantages of one,
is more practicable than in any other department of natural history ; provided a shell is perfect of its kind, and exhibits all its specific characters,
nothing more, as regards science, is requisite. There are, besides, a number
of shells, which, either from their beauty or excessive rarity, have been
figured by almost every conchological writer, and are so well known, that
their representations and descriptions thus preclude the necessity of their
purchase. Upon these principles, we formed our first collection, now in the
possession of the Manchester Natural History Society. It contained more
than 2510 species, and about 6150 specimens. On the other hand, the collection of shells made by the late earl of Tankerville was obviously formed
with a primary regard to size, beauty, and perfection, science having been
considered secondary : we know, from an authentic source, that this collection cost near 6000*l.*, while the number of species, according to the
printed catalogue, did not exceed 2487.

types of every genus of birds, or only of every family, as the case may be.   If he is particularly attached to the *Trochilidæ*, or humming-birds, and wishes to investigate them, he will procure types of the different forms observable among the other slender billed or suctorial birds (*Tenuirostres*), in order that he may study the affinities and analogies they may bear to his favourite family.   Again, the entomologist who may feel a predilection for Lepidopterous insects should satisfy himself with singling out one or two families, and directing his attention to acquiring as many individual species of these, and these only, as possible : a series of types, showing the forms of the neighbouring divisions, is all he will require to give him a correct idea of the station his favourites may hold among their congeners.   Collections, formed upon these principles, contribute, above all others, to the effectual advancement of science, while their formation, also, is comparatively easy ; the attention of their possessor is not distracted by numberless drawers of unnamed specimens, nor is he tempted to wander from the object it has previously fixed upon.

(110.)   *Typical collections* consist only of single specimens or examples of forms, either characteristic of families or of genera.   They are peculiarly calculated to give general and enlarged views of the science, but are insufficient to supply its details in the history of species.   In all public museums, and even in private general collections, a series of generic types should be kept apart, that the eye may be accustomed to the different forms, and the memory refreshed by their names.   Typical collections are more useful to a student than any others, as they exemplify those divisions which it is important should be strongly impressed on the memory, long before he descends to a knowledge of species.   The entomologists of the present day wisely endeavour to procure the greatest number of generic types (whether of British or exotic insects) in preference to a multitude of species.

(111.) *Local or geographic collections* are confined to
the animals of one particular country. They are attended with this advantage, that they can be formed
with little comparative expense ; and they acquire (in
the eyes of their possessor) an additional interest and
value from being the fruits of his own exertion. Collectors of indigenous animals are generally very particular in not admitting into their cabinets any specimens
not actually *natives,* although the species may have been
taken alive in Britain ; nevertheless, this rule, by others,
is thought to be too fastidious. However useful and
important it is to see, and be acquainted with the productions of one country, and of our own in particular,
yet their study alone is insufficient to give enlarged
conceptions on natural relations ; neither can they contribute much to a knowledge of zoological distribution.
The views of the student are circumscribed by an artificial definition in geography ; and like those who study
a subject through a confined medium, he will be in
perpetual danger of confounding local peculiarities with
general principles.

(112.) *Economic* collections are restricted to those
animals whose injurious or beneficial qualities more immediately relate to the operations of man. It is surprising
that collections of this nature are seldom if ever seen,
as they might be made not only interesting but instructive and important to those more immediately engaged
in agriculture, commerce, manufactories, and the arts.
That this may be more readily comprehended, we shall
briefly notice the principles upon which a few such collections might be formed ; the advantages to be derived
therefrom will be sufficiently obvious to agriculturists,
planters, and gardeners. A series of specimens in their
different stages of metamorphosis, of all insects beneficial
or hurtful to trees and cultivated vegetables, together
with occasional specimens of the substances they attack,
and by which the nature of the injury is at once seen ;
to each of these a label should be attached, referring to
a note-book or journal, in which is entered the time

of the insect's first appearance, the period at which it is most injurious, and the degree of success that may have attended the various operations for its destruction. The chemist and druggist, if he wishes to acquire a satisfactory knowledge of the various animal and vegetable substances belonging to his business, should endeavour to procure specimens of as many as possible in their native or unprepared state. The silk merchant, in like manner, would derive a just and sound knowledge of the various insects whose fabric constitutes his riches, by procuring specimens of the larva, pupa, and perfect moth of the silk insects, from all parts of the world ; for he may, perhaps, not be aware, that there are, in Asia alone, no less than five or six distinct kinds, which have never been exported from that country, and of whose history, to this day, we know but very little. Our knowledge is equally imperfect respecting the identity of several animals whose furs constitute an important branch of commerce with America.   The application of natural history to the practical purposes of life has been much dwelt upon by many writers in a general way ; and urged as a sort of apology (as if *any* apology was necessary) for studying the works of the Creator ; but, unfortunately, very few have hitherto applied this truth to any particular purpose.

(113.) Collections formed without reference to any general or connecting plan, are not scientific ; although they may be very useful in exhibiting the form and characters of individual species, or in calling the attention of the spectator to the beauty and variety in Nature's forms.   Cabinets of this sort are generally formed by amateur collectors for the gratification of the eye, or the decoration of the drawing-room ; yet they are not, on this account, to be despised by the scientific naturalist. In nothing has the growing taste for natural history so much manifested itself, as in the prevalent fashion of placing glass cases of beautiful birds and splendid insects on the mantel piece or the side-table.   The attention of the most indolent is attracted, the curiosity of the inquisitive awakened; and thus a first impulse may be given, particularly to youthful minds, to tastes and studies which may prove the solace and delight of after years.

(114.) The *arrangement of museums* and collections demands a distinct section in this chapter. The preservation of specimens after they have undergone the first process, mainly depends upon the security of the places where they are deposited; and their utility and beauty upon the order and taste of their arrangement.

(115.) *Quadrupeds* of a large size are seldom seen in private collections, from the great space they occupy. Those above the size of a goat may be mounted upon a thick plank, sufficiently heavy to keep the specimen firm, and either deposited on the floor in convenient situations round the museum, or placed upon the tops of other cases. Once or twice every year, each specimen should be carefully examined, and the fur and skin well saturated with spirits of wine and corrosive sublimate, in the proportion of three drachms to a pint. We can only recommend this method upon the ground of economy, for there is always reason to fear the injuries of insects to specimens not inclosed in glass cases; and where the fur or hair is long and thick, this risk is doubled. For a public museum we recommend the plan we adopted in our arrangement of the Liverpool Institution : we appropriated a low range of glazed cases, about $3\frac{3}{4}$ feet high, round the room, for quadrupeds only ; these project about a foot in front of the upper cases, and have a very neat and elegant appearance. Much space is saved by disposing monkeys and other climbing animals upon branches of trees, two or three upon each ; but these branches should be so placed as to admit of their being taken out for examination. Bats, being subjects requiring close inspection, are best kept in drawers similar to those hereafter described.

(116.) *Birds.*—The custom formerly in use in this country (and still adhered to by some), was to place each specimen in a case by itself, proportioned to its size. This plan is very advantageous on the score of security; for, if the specimen is in a sound state, the case rendered air-tight, and not opened, it will remain uninjured for a century. It is, however, objectionable in two ways: first, as occupying too much space; and secondly, as being unfavourable to a minute examination. This method, about twenty years ago, gave place to another, much more elegant and compact. For land birds, large and small branches of trees are fixed into glazed cases, upon which the birds are grouped: little or no space is thus lost; and, if disposed with taste, they have a striking and beautiful effect. The water birds, in like manner, are grouped upon artificial rocks made of brown paper sprinkled with sand, &c.; yet still the subjects, from being permanently affixed, cannot be minutely examined, and those characters distinctly seen which are essential to its scientific description. This objection we endeavoured to obviate by the following plan: — For land birds of a small size, we had cases made about 2 feet 5 inches high, 20 inches broad, and $9\frac{1}{2}$ inches deep, the front being composed of a single pane of glass: one principal stem is made to send forth smaller branches, upon which the birds are permanently fixed, as in the common method; but the main stem is strongly wedged into a square piece of lead, sufficiently weighty to support the whole; this lead is kept steady in the case by three of its sides being confined by slips of wood, or by the more secure expedient of being perforated for the reception of screws, which fasten it to the bottom of the case. When these screws are removed, the whole contents may be drawn out by a brass ring fixed in the lead, and the observer can thus handle and examine every specimen. For large birds, or those between the size of a thrush and a magpie, a different plan may be pursued. The cases are of two sizes; one being 2 feet 4 inches high, by 4 feet 2 inches broad, with three panes of glass; the other 4 feet 8 inches high, of the same breadth, with nine panes; the depth of both is 13 inches: the smallest, placed lengthways, serves as a pediment for the other: both are fitted up with strong branches, firmly secured by screws to the back and sides of the case. Each bird is mounted upon a short strong stick, one end of which is made round, and is thrust into a corresponding hole in some of the branches : they should then be so firmly secured, that the bird is supported without any other assistance ; but if it be large, or unusually heavy, a little glue may be added. Now, the advantages of this plan are great ; for while the general beauty and variety of the whole is preserved, the naturalist is able to detach any one particular species for the purposes of examination, and to replace it, without the slightest injury to any others that may be placed upon the same branch.

(117.) Birds in public museums, however, should be mounted on small wooden stands, and placed upon flat shelves within glazed cases : such as are of a large size, and all those which live in the water, are more advantageously put upon flat stands, made of wood painted white, or plaster of Paris, in which holes are made for the passage of the wires. This method, which we have borrowed from the French, will not admit of displaying much beauty or taste, or of giving to the collection any great effect ; but it combines two important advantages, for the specimens, being isolated, may be removed or rearranged, and the student can examine each without the least trouble.

(118.) The preservation of *birds in skins*, or, more properly, in an unmounted state, is, above all others, the best for scientific purposes. Unless a bird is mounted more carefully than is usually done, some part, either of the bill, nostrils, toes, or claws, will be injured or distorted : it is, in fact, very rare to see exotic birds, after they have come from the hands of the bird-stuffer, in a thoroughly perfect state. Mounted specimens, even with the most ingenious contrivances, occupy a vast deal of room ; and their preparation, at all times, is attended with much expense. Now, all these objections are removed by preserving them, as it is termed, in skins : when laid upon fine cotton, and arranged in cabinet drawers, they have a very pleasing appearance ; they

can be at all times handled, and minutely examined,
without the least trouble ; moreover, they lay in such a
compact space, that, in a cabinet $5\frac{1}{4}$ feet high, 3 feet 3
inches broad, and 1 foot 7 inches deep, containing 36
drawers, we have a collection of near 600 specimens.
Birds that have been purchased at sales, or otherwise,
and that are already mounted, we dispose in cases upon
moveable stands, as before described. Until proper cabinets are provided, bird skins are preserved with great
additional security, by wrapping them singly in sheets
of soft paper, and examining each once or twice a year.

(119.) *Bell glasses* are much used for select specimens, and are very advantageous for small creepers,
humming-birds, and others with richly coloured plumage : the bottom of the glass should fit into a corresponding groove in the stand, and not be cemented
down, as a minute inspection of the objects is thereby
prevented. In the arrangement of a collection of birds,
as well as of all other animals, a series of generic types
may advantageously be kept distinct from the principal
collection. The distinguishing forms of nature are thus
brought into immediate contact ; being at once placed
before the eye of the spectator, their variations will be
more obvious, and more strongly impressed upon the
memory. The generic and specific, together with the
English name, may be written upon a card, and affixed
to such specimens as are mounted on stands or kept in
skins. For those disposed upon branches, a small number placed near each, and referring to a catalogue, is
most preferable : this remark is equally applicable to all
other objects so disposed, excepting shells and insects.

(120.) The preservation of birds, even when deposited in the museum, requires constant watchfulness.
The moment the feathers of a specimen appear discom-
posed, falling off, or any dust is observed at the bottom
of the case, its contents should immediately be examined,
and the infected specimens removed. This, if done
immediately, will frequently stop the progress of con-
tagion. We are, at this moment, suffering from this

evil : those specimens which the moths have attacked,
have had their plumage well saturated with spirits of
wine and corrosive sublimate, in the proportion of $2\frac{1}{2}$
drachms to a pint : this, however, has been found too
weak; for, five days after the application, the young larva,
scarcely $\frac{1}{10}$th of an inch long, and inclosed in their
cases, have been found devouring feathers steeped in
this strong poison ; the corrosive sublimate has therefore been increased to 4 drachms : in a few days all
the specimens will be placed in an oven sufficiently
heated to kill the insects, yet not to crisp the feathers ;
they will then be placed in a close box for two or three
months ; and should the insects by that time appear
totally eradicated, the specimens will again be restored
to their former situation. Skins should be treated in
the same way, and invariably kept distinct from the
collection, until the destruction of the insects is well
ascertained.* Notwithstanding, however, the efficacy,
in most instances, of the fumes of camphor, spirits of
turpentine, &c., when inclosed in cases infected by insects, we give the preference to prussic acid, as the
most destructive poison to the insect world ; three or
four drops, upon a piece of cotton, will be sufficient for
a moderate drawer containing birds or insects, and about
double that quantity will be equally efficacious for a
large case. In both instances, the seams or openings
should be rendered air-tight, that the fumes may not immediately escape. Our friend, Dr. Trail, recommends
a bladder filled with oil of turpentine for large cases of
birds : a remedy we have also used with success.

(121.) The *situation* of the museum should be perfectly dry. The rays of the sun, in process of time,
materially injure the colour of birds, and, indeed, of all
similar objects; the admission of too much light into
the room should, therefore, be carefully avoided. A fine
collection of insects, formerly in the Leverian Museum,
by being exposed in glass cases round the room, was
totally spoiled in a few years.

* This passage was written some years ago.

(122.) Fish, reptiles, and amphibians, &c., preserved in spirits, are arranged on narrow shelves, and protected in front by one or two wires ; the bottles best adapted for this purpose are made with a wide base, which prevents them from being easily overturned. Fish, serpents, &c., prepared in a dry state, are generally put into cases, and disposed upon artificial rock-work ; but this, although it may please the eye, renders them useless as objects of study. The admirable disposal of these animals in the museum of the College of Surgeons is a model for all others.

(123.) INSECTS, in regard to their arrangement, are disposed either in corked boxes or in glazed drawers. Most of the foreign entomologists choose the former, on account of the great expense attending the construction of glazed cabinets, which, if of the size usually preferred by our own collectors, average near a guinea a drawer. M. Samouelle observes*, that a collection of British insects requires a cabinet of from 50 to 100 drawers, 14 or 15 inches square. We do not, however, think the advantage of a superior display is worth paying so dearly for ; and shall therefore recommend young entomologists to deposit their collections in what are called store boxes; where their specimens, with care, will be preserved equally well, and at one fifth of the expense. If this plan is objected to for *all* the orders, the *Lepidoptera*, from their more delicate texture, may be placed under glass. Most of the store boxes we are in the habit of using, are $13\frac{3}{4}$ inches by 9 or $9\frac{1}{2}$; they open in the middle, where they are rabbeted, and in such a way, that, when laid open, both sides lie flat upon the table, each half having a depth of $\frac{3}{4}$ in the clear. When it is shut, the depth of the box outside is $2\frac{5}{8}$ inches ; this depth being neither too deep or too shallow. The sizes, however, may vary. Some of ours, adapted for large winged insects, as the locusts, dragon-flies, &c., are 13 inches by $11\frac{1}{2}$: others, for the *Diptera*, are 12 by $8\frac{1}{4}$, and only $2\frac{1}{4}$ deep outside : the

* Entomologist's Compendium, p. 310.

wood is deal, and never more than $\frac{3}{8}$ thick. A small piece of camphor, inclosed in muslin, can be fastened by a pin in one corner. This method we prefer to a wooden partition at one end, which is but a clumsy contrivance, for the lid is constantly apt to fall out when the boxes are placed perpendicularly. When neatly finished with cloth backs, and labelled, the whole may be arranged like books upon shelves, and thus have a very pleasing appearance. Both the inner surfaces are corked, and covered with white paper. By using both sides, as many insects can be placed in one of these boxes as would fill two glazed drawers of the same size.

(124.) Upon *corking of boxes* or drawers, Mr. Samouelle says : — " The readiest way is to buy the cork prepared, which may be obtained at most of the cork-cutters ; but this is expensive. I have generally bought it in the rough state, and cut it into strips about 3 inches wide : the length is immaterial, if the method advised hereafter is pursued. These strips must be fixed in a vice, and, if the substance of the cork will admit, split down the middle with a fine saw : (greasing the saw must be avoided as much as possible, as it will stain the paper used for covering it afterwards :) the outer or black side is to be rasped down to a certain smoothness, as well as the middle or inside. Having reduced the slips to about $\frac{3}{8}$ of an inch in thickness, glue each piece (the darkest or worst side) on a sheet of brown or cartridge paper. This should be laid on a deal board about 3 feet in length, and of the same width as is required for the drawer or box : a few fine nails or brads must be driven through each piece of cork, to keep it firm, until the glue is dried. By this means, sheets of cork may be formed of the size of the drawer. All the irregularities must be filed or rasped down quite even, and the whole surface rendered perfectly smooth by rubbing it over with pumice-stone. The sheet, thus formed and finished, must be glued into the drawers. To prevent its warping, some weights should be equally distributed over the cork, that it may adhere firmly to the bottom of the drawer: when quite dry, the weights are removed, and the cork covered with paper, which should be of the finest quality, but not very stout.   The paste should soak well into the paper previous to being laid over the cork, which, if smoothly laid on and gently rubbed over with a clean cloth, will be rendered smooth and tight when dry." *

(125.) Cabinets for insects are fitted up with drawers, the bottoms of which are lined, as just stated, with cork, and papered; the whole being covered with a glass frame or top, which is rendered air-tight, yet capable of being lifted off and on without difficulty.   The size of these drawers is a matter of some consideration. Mr. Kirby recommends them to be 18 inches square; yet a perfectly square form is, perhaps, not so advantageous on many accounts, as an oblong square; and the glass, of such dimensions, is proportionably more expensive. Samouelle observes, that the drawers may be from 14 to 15 inches square; but this size, although well adapted for British insects, is certainly too small for the majority of exotic species, particularly the lepidopterous tribes. After much consideration, and balancing of estimates as to cost, we had our own made $18\frac{1}{4}$ inches by $16\frac{3}{8}$ in the full, the frame of the glass being $\frac{3}{4}$ broad. The full depth of each drawer is $2\frac{1}{2}$ inches outside, and $1\frac{1}{8}$ inside, in the clear; that is, after being corked. Within, but hid by the glazed frame, is a very narrow partition parallel to the interior of the front side; this is for the purpose of containing camphor, and is perforated in all parts with small holes, that the fumes may escape. Some collectors have these partitions on all the sides; but this appears unnecessary, and occupies too much space. As some few tribes of insects among the *Coleoptera* require, from their gigantic size, a greater height, two or three drawers may be made two inches deep.   The number of drawers requisite, and consequently the size of the cabinet, depend of course upon the extent of the

* Entomologist's Compendium, p. 311.

collection, or of the plan upon which its possessor intends
to proceed. To preserve uniformity in this respect, it is
better to have each cabinet made uniformly with the
others, and of such a height that, if necessary, they may
form a kind of pedestal for another series of drawers,
somewhat smaller, and constituting an upper part; when
both these are placed together, the top drawers of the
upper division should not, for convenience, be beyond
the reach of the hand. The cabinets made by the London workman generally contain forty drawers, arranged
in two tiers, and protected by folding doors of plain mahogany. This does not appear to us in good taste, for
the whole immediately reminds one of a tent bedstead.
Doors, also, made entirely of wood, give a heavy and
wardrobe-like appearance. Those in our principal cabinet have the centre covered with rich purple silk de-
fended by brass wire. The effect this produces on the
eye is vastly superior, and the expense is rather less.

(126.) The *wood* best adapted for cabinet drawers
is mahogany or wainscot, but a combination of both in
the inside work is perhaps the best; the drawers may
be made of wainscot, and the outside of mahogany or
other dark ornamental wood. Cedar should on no account be used; it exudes a resinous gum, which collects,
not only on the wood itself, but on the objects contained
within, and inevitably spoils them. Whatever description of wood is used, it must be old and well seasoned.
From carelessness in this respect, we have seen cabinets
that have cost large sums, completely spoiled. The entomologist will see some of the best made cabinets for
insects in the British Museum.

(127.) *Drawers for bird skins* should be 20 inches
by 23, and varying from $1\frac{3}{4}$ to 3 inches in depth: in
most other respects, except in not requiring either corking or glazing, they may be made on the same principles
as those for insect cabinets. Sometimes a small oblong
space is let in at the front of each drawer, for inserting
a piece of white card or ivory upon which the contents
are written. In large collections, this is very useful;

but the same object may be gained by pasting a neat label on the outside of each drawer.

(128.) When the *insect drawers* are corked, and the bottom and sides neatly covered with paper, the collector has next to *arrange* his specimens. The usual method of doing this is, to rule straight pencil lines parallel with the sides, and to place the insects between: they will thus be disposed in columns, the female following the male; and the series of species or genera continued from the termination of one column to the commencement of the next. The space between these lines will, of course, be regulated by the size of the insects; and the separation of genera is signified by double instead of single lines. A label bearing the generic name is placed at the head of each group. The first insect should be typical; after which we place the aberrant species allied to the preceding group; then the remainder of such as may be considered typical; and lastly, those which are again aberrant, and lead to the following genus. If the collector intends to increase his cabinet, vacant spaces should be left, at discretion, for desiderata. Under each species, when ascertained, should be affixed the specific name, written or printed upon a narrow slip of card or paper, and secured by a pin: this is better than putting numbers referring to a catalogue; for every method by which names may be impressed upon the memory, should be adopted.

(129.) *Dissections* of the head, antennæ, wings, &c., if displayed on cards, tend much to increase the scientific value of collections; and they may be made from such duplicates of typical forms, as may not be wanted for perfect specimens. The parts should be neatly affixed upon a slip of card, and placed, with another example of the insect in an entire state, at the head of the genus. Thus, to illustrate the genus *Pieris*, the common garden butterfly (*P. brassica*) may be selected, and the following parts displayed : — The head, with the mouth uppermost; the outer and inner sides of the palpi; the antennæ, as seen on both sides; the fore, middle, and hind

legs ; and lastly, the abdomen of the male and of the
female. By this plan, the distinctions between different
forms are at once made prominent, and tedious examin-
ations (frequently necessary when the naturalist is most
in haste) in a great degree prevented. The more instructive an arranged collection can be made, the more
scientific and useful it becomes. The present methods
now in use, in this respect, are capable of much improvement. In a general collection of insects (more particularly if it belongs to a public museum, where instruction
should be the primary object), it would be desirable that
all the divisions of orders, tribes, families, and sub-
families, should be exhibited to the spectator at one
view, by a typical example of each, placed close to its
approximating form, and in its proper line of affinity.
Thus, in the first drawer of the cabinet, may be placed
a series of specimens exemplifying the orders into which
all insects have been divided. If the circular system is
adopted, two circles can be drawn with pencil, — one for
the *Aptera*, the other for the *Ptilota* : the specimens of
each order are then placed at proper distances on the
pencil lines. At the head of each *order*, a similar disposition should be made of the types which represent
the *tribes* composing the order ; next will follow examples of the *families;* and thus, gradually leaving one
group for an inferior one, we descend to the genera and
species, which may be arranged upon the plan already
proposed.

(130.) *Kirby and Spence's plan* of arranging insects
in cabinets is thus described : — " Divide each drawer
transversely by a full black line; parallel with this, on
each side, draw a line with red ink ; then, for arranging
your insects, draw pencil lines, which are easily oblite-
rated, at right angles with the others, according to the
general size of the insects that are to occupy them.
Insects look better thus arranged in double columns,
than if the pencil lines traversed the whole width of the
drawers. In arranging them, you may either place them
in a straight line *between* the pencil lines, — which I

think is best, — or upon them. You will begin your columns from the red lines in the middle, and not from the sides of the drawer : thus, the heads of those on one side of it will be in an opposite direction to those on the other. Where your pins are very fine and weak, you must make a hole first with a common lace-pin ; otherwise, in forcing them into the cork, they will bend. In labelling your specimens, you should stick the appellation of the genus or sub-genus with a pin before the species which belongs to it."*

(131.) *Camphor* is necessary to preserve the specimens from the attacks of *Acari*, and other minute insects, whose presence is easily detected by the appearance of dust in the drawers. Mr. Kirby observes, " Some insects, in a chip box, having become much infected by *Acari* and *Psocus pulsatorius*, I placed, under a wine glass, several of each, along with roughly powdered camphor : at the end of twenty hours, the *Acari* were alive ; but at the end of forty-eight, they were all apparently dead, and did not revive upon the removal of the camphor. The specimens of *Psocus* all appeared dead in an hour, and never revived. If the camphor be put only into one side of a drawer, and in a lump, though, perhaps, it may keep out *Acari*, &c., it will not expel them. When any specimens appear — from the quantity of dust that may be observed round them — to be very much infected by these insects, the under parts of the head, thorax, and body should be anointed with spirits of wine and corrosive sublimate (3 drachms to a pint), applied with a camel-hair pencil, and the specimen suffered to dry before it is again replaced in the drawer. Some of the German naturalists, when in Brazil, anointed all their insects in this way with a weak solution of arsenic soap, which they considered the most efficacious preservative against the attack of ants; but this soils the under side, and almost always leaves the specimens moist. Large coleopterous insects, of a black colour, may be preserved from being devoured, by

* Kirby and Spence, iv. p. 545.

slightly washing the upper and under parts with a brush, dipped in well rectified spirits of turpentine; but this is very apt to change and dim the colours of many insects, and should be used, for *Lepidoptera*, with great caution. After all, we give a decided preference to the use of prussic acid, two or three drops of which, dropped into a drawer, is a sure method of destroying all the devourers of insect specimens. The drawer should, of course, be immediately closed as tight as possible.

(132.) SHELLS are generally arranged in cabinets, the drawers of which are merely whitened or papered inside. The most convenient size for these will be from 17 to 18 inches square: their depth should be various; thus, we should say that two thirds of any given number of drawers in a cabinet should be 2 inches deep, one half of the remaining portion about $1\frac{1}{2}$, and the rest $2\frac{1}{2}$ inches. The generality of collectors display their specimens upon a piece of fine carded cotton (such as is used by jewellers), cut to the size of each drawer: this is a very good way of showing them to advantage, but every movement disturbs their position; besides, small specimens get entangled in the cotton, and are frequently lost, unless put into little card boxes (either round or square), which the collector may buy or make for himself. Many specimens will be too large for the generality of drawers, and are therefore disposed in groups upon the outside of the cabinet, or protected by glass cases. The French method is to affix the specimens upon square or oblong slips of thin wainscot, of various sizes, but so proportioned that they will all fit close to each other, without leaving any intermediate space so small as not to be filled up by any one of the sizes. The upper surface of these tablets is covered with white paper, and one is appropriated to each species; the specimens are affixed by strong gum-water, and the generic and specific name written underneath. If the collector possesses more than one example of a species, the other is affixed in a different position, in order that the upper and under surface may be shown: in like manner, a series of young specimens is placed in their progressive order of growth. Black paper may be substituted for white, for such small or colourless shells as require a strong contrast. This is, perhaps, the best method of arranging the generality of shells in a public museum, as, when once fixed upon the boards, they can neither be injured nor displaced; but it is not well adapted for shells above a certain size. The large and beautiful volutes, cones, &c. are therefore best preserved upon cotton, in glass cases, or in drawers.

(133.) *Card trays* are used by many collectors, who object to boards, as being expensive and cumbrous: hence are substituted square boxes, made of cards in the following manner: — Parallel to the four sides of the card, a straight line is cut by the point of a penknife, sufficiently deep to admit of one half of its substance being cut through, and folded back without difficulty: the space between the edge and the cut line will, of course, constitute the depth of the box, and may be varied according to the fancy of the collector, or the nature of the specimens it is to hold: when these four sides are cut, the corresponding corners are taken out by the scissors, and the sides bent up and united by pasted slips of paper. The bottom of the box may afterwards be covered either with black paper or velvet, and the specimen placed within. It is better to affix small or minute shells with gum water, or all the specimens of a moderate size may be secured in the same way. A series of generic types should be arranged in the first drawers, and the indigenous species may be distinguished by their names being written upon pink-coloured paper.

(134.) *Crustaceous animals,* if not of two large a size, are very conveniently placed within card boxes; but they should not be fastened down, as the under parts sometimes require to be examined. The smaller species may be stuck through with a pin; or, if both surfaces can be exhibited, the specimens may then be secured with gum. *Echini,* and other marine productions of a similar nature, may be placed either upon cotton or within card boxes.

(135.) *Corals,* from their size and fragility, require the space and protection of glass cases. The base of each specimen should be firmly affixed to a stand, made either of plaster or wood, proportionate to the size and weight of the subject it is intended to support. For wooden stands, black is the best colour ; the name should be affixed to each ; and they may be arranged on horizontal shelves.

(136.) *Catalogues* of the subjects contained in all public museums should not only be drawn up, but printed and circulated, the new species shortly described, and a list of desiderata subjoined. Information is thus communicated to naturalists in all parts of the world, and their assistance solicited towards the acquisition of new objects. In ordinary cases, such catalogues need only contain the generic and specific name; but a reference should be made to some author where the species has been figured or described.

(137.) In concluding this part of our volume, there is one caution we feel bound to mention, or we may unintentionally mislead. We have occasionally used the expressions important and valuable; but these terms are only applied to subjects in natural history in a scientific, not in a commercial, sense. To the naturalist who collects merely for his individual amusement or instruction, the marketable value of his acquisitions is never thought of ; or it is, at least, of no consequence : he can give away his duplicates, or exchange them with his friends. But we warn the commercial naturalist against indulging any hope, that the sale of his collections, even if he explore foreign countries, will at all remunerate him for the trouble, the anxiety, and the expense of their acquisition. If he collects on account of others, it is all very well ; but he will be sure of disappointment, if he expects " the public " will give him such prices as will render it worth his taking the risk of remuneration upon himself. It is now a well-known fact, that collections of South American insects can be purchased cheaper in London than in the country they actually inhabit. Birds are still more saleable here. We have purchased, in fact, hundreds below the average price that their mere stuffing would cost in England. Shells are now a complete " drug on the market." The *Helix hœmastoma,* which used to sell, at auctions, for eight or ten shillings, can now be had for — sixpence! and North American unios, or river bivalve shells, have been actually sent to a well-known London dealer in hogsheads, although a few years ago specimens, now not worth a shilling, sold readily at fifteen or twenty. Let no one, therefore, think of collecting for *profit*, or it is ten to one he finds himself a heavy loser. We should recommend all collectors, who have duplicates, either to exchange them with their friends, or to give them as contributions to the local and provincial museums, now forming in almost every town in England: better to do this than make them an unwilling present to " the scientific public," by sacrificing them at an auction.

H

PART II.

A

BIBLIOGRAPHY OF ZOOLOGY;

WITH

BIOGRAPHICAL SKETCHES

OF

THE PRINCIPAL AUTHORS.

———————

WITH all our desire to introduce, into this part of our
volume, every important work on General Zoology, and
every Author by whom the science has been advanced,
we are fearful it may be defective in a few instances.
Some publications, of very recent date, have not yet
come to us from the Continent; while others, from the
diffuse nature of the subject, may have escaped our research: the American works, more especially, are most
imperfectly known in England; where, in fact, they
cannot be purchased. With all these obstacles, however, we believe our list will be tolerably perfect, since it
is much more extensive than any we have yet met with
on general zoology. Although it considerably exceeds
that of the *Règne Animal*, there are several instances
wherein our information is entirely derived from that
admirable basis of zoological bibliography.

———————

Abbot, John. — *Entomology.*

A most assiduous collector, and an admirable draftsman of insects. At an early age he was engaged by three or four of the leading entomologists of England, to go out to North America, for the purpose of collecting insects for their cabinets. After visiting several parts of the Union, he determined to settle in the province of Georgia, where he immediately began his researches. The late Mr. Francillon, whose magnificent collection of insects, which rivalled that of Drury, is still remembered, was his chief friend and correspondent, through whose means and agency he procured large commissions from the British and Continental collectors, and different public museums, for Georgian insects. Abbot's specimens were certainly the finest that have ever been transmitted as articles of commerce to this country: they were always sent home expanded, even the most minute ; and he was so watchful and indefatigable in his researches, that he contrived to breed nearly the whole of the *Lepidoptera.* His general price, for a box-full, was sixpence each specimen ; which was certainly not too much, considering the beauty and high perfection of all the individuals. Abbot, however, was not a mere collector. Every moment of time he could possibly devote from his field researches, was employed in making finished drawings of the larva, pupa, and perfect insect of every lepidopterous species, as well as of the plant upon which it fed. These drawings are so beautifully chaste and wonderfully correct, that they were coveted by every one. So many, in fact, applied for them, both in Europe and America, that he found it expedient to employ one or two assistants, whose copies he retouched ; and, thus finished, they generally pass as his own. To an experienced eye, however, the originals of the master are readily distinguished. Mr. Francillon possessed many hundreds ; but we know not into whose hands they have now passed. Another series of 103 subjects, not included in that which has been pub-

lished, was executed for us, with the intention of forming two additional volumes to those edited by Dr.
Smith : but the design is now abandoned. The healthy
and peaceful occupations of this meritorious entomologist has led to great length of life ; for we had the
pleasure of receiving a collection of insects from him
only two years ago.   He is probably now above 80.

The *Natural History* of the rarer Lepidopterous
Insects of Georgia, collected from the Drawings and
Observations of Mr. John Abbot.   Edited by Sir
James Ed. Smith, M.D.   2 vols. folio.   London,
1797.   With 104 coloured plates.   Having already
spoken more particularly of this noble work in another
place*, we can only repeat, that it is one of the most
beautiful and valuable that this or any other country
can boast of.   There are many inferior copies on sale
among the booksellers, which are offered at a low
price, but the original coloured impressions are seldom met with.

ACERBI, JOSEPH. — *Zoology and Botany.*

An accomplished Swede, and observing traveller, in
whose writings are many valuable remarks on natural
history.

Travels through Sweden, Finland, and Lapland, to
the North Cape, in 1798 and 1799.   London, 1802.
2 vols. 4to. with numerous plates, many of natural
history.

ADANSON, MICHEL. — *Traveller and General Naturalist.*

This eminent naturalist and traveller, although properly associated with the science of France, was originally,
as his name would even imply, of Scotch extraction ; although the cause which induced his parents, who appear
to have been wealthy, to settle in France, is unknown.

* Preliminary Discourse, p. 66.

He himself was born at Aix, in Provence, on the 7th
of April, 1727, and began at a very early age to show
indications of unusual talent. He was sent to pursue
his studies at the university of Paris, where, it appears,
he first imbibed a taste for natural history, in consequence of being presented with a small microscope. The
latent spark thus kindled, he devoted all the time he
could spare from the usual studies of the university, to
the instructions and lectures of Réaumur in zoology,
and the great Bernard Jussieu in botany. These so
inflamed his youthful imagination, that he soon determined to decline the professional pursuits, arranged for
him by his parents, for the more animating and congenial object of travelling in unknown regions, and investigating their productions. Accordingly, at the early
age of 21, he arranged his plans, and determined, at his
own expense, to visit the little known colonies of the
French and other European nations on the coast of
Western Africa, regions which had hitherto never been
visited by the naturalist. He embarked for Senegal in
the year 1748, and, after visiting the Azores and Canary
Islands, landed on the island of Goree. In this situation,
admirably adapted both for botanical investigations on
the main land, and for prosecuting the study of the marine
mollusks which swarmed upon the shores, he remained
for five years, incessantly employed in investigating
and describing the natural productions collected around,
as well as extending his observations on the physical
peculiarities of the country and its native tribes. He
returned to France, loaded with these most valuable
treasures to science, in 1756, and immediately began to
arrange them for publication. The same year he was
elected a corresponding Member of the Academy of
Sciences, — an honour he showed was not undeserved, by
communicating two interesting botanical essays. In the
following year appeared the first quarto volume of his
Natural History of Senegal, prefaced by a short account
of his voyage, but chiefly occupied by his account of the
testaceous mollusks, whose animals he had drawn and

described on the spot. A poor abridgment of this narrative, badly executed, was published in London in 1759. In 1763, appeared his *Familles des Plantes*, in two octavo volumes,—a most important work, which would have created, even at the time, a much greater sensation in the botanical world, but for the great and paramount influence of the Linnæan artificial system. Nevertheless, its merits, at once seen by the discerning few, gradually became so generally admitted, that a second edition was ultimately given to the public. The enthusiastic spirit of Adanson, however, seems to have conquered his better judgment; for, instead of confining his attention, in the first instance, to making known his African discoveries, he aimed at accomplishing a task altogether superhuman : this was nothing less than publishing a complete Encyclopedia of Natural History. So strongly was his mind bent upon this impracticable project, that he laboured upon it for many years ; and it clung to him even when laid upon that bed from which he never rose. His materials for this gigantic undertaking were immense ; and he expended the emoluments of the various offices he held under the crown, as Academician and Royal Censor, in augmenting them from all quarters of the globe. Yet Adanson was not a mere naturalist, or a visionary projector. His reputation stood so high, as a man of sound judgment and commanding talent, that, in 1760, the British government made him liberal offers for communicating to them his plan for forming a regular colony on the coast of Senegal (then in possession of this country), in which he proposed to cultivate its productions by the free labour of the natives. This honour he wished to see acquired by the French ; and the same patriotic feelings which induced him to decline the overtures of the British government, led him also to refuse the liberal offers successively made to him by the emperor of Germany, Catherine of Russia, and the king of Spain, to reside in their dominions, and add lustre to their courts. He was too much attached to his native country, to accede to

any proposal that would thus draw him away from his
numerous friends, and interrupt the course of his pursuits. He continued, therefore, to reside in France, notwithstanding the troubles which began to distract the
nation and unhinge society: but the revolution soon
followed, and stripped him of every thing. To such poverty, indeed, did this fearful convulsion reduce him,
that, being subsequently invited to become a member of
the National Institute, he replied, that he could not
accept it, "having no shoes." The minister of the
interior, much to his honour, did his utmost to rescue
the noble Adanson from the horrors of poverty: he
succeeded in procuring him a pension, sufficient to support him, in moderate independence, until the time of
his death. This event happened on the 3d of August,
1806, in the 79th year of his age, after an illness of several months, which confined him to his bed. Whether
his numerous collections and materials, for completing
his African researches, and for his "Encyclopedia of
Nature," were destroyed in the revolution, or were dis-
persed by sale, is unknown : but it appears certain he
published nothing of consequence after his "Families
of Plants." By grasping at too much, he accomplished
too little, even for a man of ordinary application. How
frequently do we find this is the only result of an enthusiastic temperament, an inordinate passion for science,
and excursive talents. In his private life and feelings,
Adanson is said to have had many excellent qualities ;
he was indefatigable in his studies, but somewhat con-
ceited ; and very careless withal, both in his dress and
manners,— small defects, indeed, but which are much
against a man in the every day transactions and inter-
courses of life. He seems to have had a violent and
even ridiculous feeling of hostility towards Linnæus,—
a circumstance which militates against assigning to him
greatness of mind. Linnæus bore his enmity with philosophic calmness, and more than once said that he believed Adanson was either mad or intoxicated. Whether
the opinion of Haller was well founded, that these two

great men were so equal in talents, as to be worthy
rivals of each other, we cannot pretend to decide. Their
powers of mind, indeed, might have been equal, but
they had not the same temperament. The aim of both
was the same,—that of being a commentator upon all
nature, as then known. But the one, by the unrivalled
system of nomenclature he pursued, lived to see almost
the end of his stupendous task; while the other, pursuing the excursive and more tedious plan of Buffon,
and hindered by adverse circumstances, accomplished
comparatively little or nothing. The life of Adanson
is instructive; since it shows the wisdom of confining
our studies within moderate limits,—without which the
highest talents are almost useless, and the greatest
energies of mind become expended in vain.

   Histoire Naturelle du Sénégal. Coquillages. Paris,
   1757. 4to. pp. 275. pl. 19. The animals of the
   shellfish, drawn and described on the spot, render
   this book particularly valuable, although the figures
   are not above mediocrity.

AGASSIZ, L.—*Ichthyology.*

A well-known German ichthyologist. Besides many
papers in the *Isis*, and other periodical journals of the
Continent, but which are almost unknown in this
country, M. Agassiz has acquired celebrity by his investigations of fossil fishes; while his masterly descriptions, interspersed with those of Spix, place him
in the foremost ranks of modern ichthyologists. He
long ago announced a work, with coloured figures, on
the freshwater fishes of Europe; but this we have not
yet seen.

   Selecta Genera et Species Piscium Brasiliensium:
   digessit, descripsit, et Observationibus illustravit
   L. Agassiz; præfatus est, et edidit C. F. P. von Martius. Monachi, 1829—31. 2 vols. large 4to. 101
   coloured plates.

AHRENSIUS, AUGUSTUS. — *Entomology.*

Fauna Insectorum Europæ, curâ E. F. Germar, et Fr. Kaulfuss. Halæ, 1812—23. Published in 12 oblong 8vo numbers, each with 25* coloured plates.

ALBIN, ELEAZER. — *Zoological Artist.*

A professional painter, cotemporary with George Edwards, but without much knowledge of natural history, and a very indifferent artist.

1. *British Birds*, a Natural History of. Illustrated with copperplates, curiously engraven from the life, by E. Albin, and coloured by his daughter and self. Lond. 1731—8. 3 vols. 4to. 306 coloured plates.

2. *English Song Birds*, a Natural History of. London, 1759. 8vo. pp. 96. 306 coloured plates.

3. *Spiders*, and other Curious Insects, Natural History of. London, 1736. 1 vol. 4to. 53 coloured plates.

4. *English Insects*, Natural History of; with Notes and Observations of W. Derham. London, 1749. 1 vol. 4to. 100 coloured plates.

5. *Esculent Fish*, Natural History of; with an Essay on the Breeding of Fish, and the Construction of Fish-ponds. By the Hon. Roger North. 1 vol. 4to. 18 coloured plates. London, 1794.

ALBINUS, BER. S. — *Comparative Anatomy.*

He was professor at Leyden, and one of the great anatomists of the eighteenth century. Born in 1697; died in 1770. He contributed several valuable essays to the

Annotationes Academicæ. Leyden, 1754—1768. 8 Nos. in 4to.

* Percheron says, " 24 figures in each, of different orders."

ALDROVANDI, ULYSSES.—*General Zoology.*

Respecting the leading circumstances of the life of
Ulysses Aldrovandi, who has long and familiarly been
known as " The Modern Pliny," and who, upon no less
authority than that of Haller, is reckoned " the most
skilful naturalist of his time," it might reasonably have
been expected that nothing like obscurity or doubt
should have remained ; and yet we venture to affirm,
after taking some pains in the matter, that this is really
the case.   Aldrovandi's extraction was noble ; being
descended, according to his contemporary Miræus, from
counts of that name, and, on his own, the best possible
testimony, from the famous Lombard general Hilde-
brand.   The exact time of his birth is doubtful, being
given by his several biographers in the years 1522,
1525, and 1527 ; and little is generally recorded of him
till he took his medical degree in his native city, Bologna, somewhere about the age of thirty.   From his
own pen, however, we learn, that in his early years he
commenced with the politer studies; that he then devoted himself for seven successive years to the civil and
canon law, and was urged to adopt these as his profession ; that he then tasted the elements of philosophy
and logic; and finally, spared no labour to make himself
acquainted with natural history, as that study which is
accompanied with the most exquisite gratification and
astonishment.

Upon taking his degree in 1553, Aldrovandi settled
as a physician in Bologna; and, according to Bullart,
taught physic within the walls of its university.   In
the year 1554, he was appointed to the chair of Philosophy and Logic, and also to the lectureship of Botany.
Besides these, as he expressly states, he was Professor
of Natural History.   Whether he filled more than one
of these chairs at a time, and in what order he received
and resigned them, no where appears; but the last was
his favourite, which he never would forego: and upon his

unwearied labours as a lecturer for a period of more
than half a century, by himself and his numerous and
celebrated associates, at home and abroad, in the fields
and his study, in the dissecting room and museum, we
must forbear to dwell.   We remark, however, that in
Bologna he left two striking monuments of his energy
and power.   The one of these was its Botanic Garden,
of which he was the founder; and so zealous a cultivator was he of this science generally, that, 100 years
after, sixteen great volumes of his *Hortus Siccus* pro-
claimed his industry and skill.   The other, and still
more striking memorial of his energy, was the Natural
History Museum, containing specimens in all departments of the science, numerous drawings illustrating
these, together with paintings and sculptures.   All these
he bequeathed to the Senate ; and many, we doubt not,
still remain to tell the tale.   Of the value of the pictures, some notion may be formed, when we state, that
Napoleon laid the spoiler's hand upon many of them
and removed them to Paris, whence they were restored
at the peace to their proper resting place.   But by far
the most substantial monument of the Bologna professor was his WORKS, now comprised in thirteen pon-
derous folios, — in fact, an immense Encyclopædia of
Science of the 16th century, one of the first of its kind,
and fully sustaining Bayle's character of its author,
" the most inquisitive man in the world with regard to
natural history."   His title to the authorship of these
Opera has been, and now generally is, impugned ; but
this arises mainly from a blunder of the Abbé Gallois,
which has been partly adopted by Bayle and others.
The injurious aspersion has already been rebutted; and
its falsity might, we believe, without difficulty, be un-
answerably demonstrated.   Six of the tomes were published during the author's life, and were by no means
those which he contemplated would first see the light.
From various causes, the last did not appear till half a
century after his death.

We must not omit to add, that Aldrovandi was a

traveller; and when, to the expense which this involves, we add those incurred in purchasing rare objects in every department of natural history, in carefully preparing these, in depicting them, with the help of first-rate masters, and engraving these drawings by the hands of first-rate artists, and getting them cut in wood by those who were renowned in this department, and finally, in printing and publishing them in a style truly sumptuous, we can readily understand how all this should involve the professor in an expenditure far beyond what his private fortune could sustain. Hence he was no stranger to pecuniary difficulties; and on this fact has been reared the story, that he died not only a poor man, but a miserable pauper in a wretched workhouse. With Haller and Cuvier, we do not subscribe to this opinion. Amidst all his difficulties, he had liberal patrons, of whom we name the popes Clement VIII. and Sixtus VI., and, most of all, the Senate of his native city, which not only liberally supplied his own exigencies, but subsequently contributed largely to the publication of his works. Towards the close of his life he became blind as well as poor; but there is reason to conclude that the interesting old man died contented and happy, in the consciousness that he at length had executed the task he had undertaken, (to use his own words) of rearing in his WORKS, to the Divine Author of Nature, some memorial, however insignificant, of a grateful heart.—J. D.*

1. *Aldrovandi Opera Omnia.*—Ornithologia, 3 vols.; De Insectis, 1 vol.; De reliquis Animalibus exsanguibus, 1 vol.; De Piscibus et Cetis, 1 vol.; De Quadrupedibus, 3 vols.; De Serpentibus et Draconibus, 1 vol.; De Monstris, 1 vol.; Museum Metallicum, 1 vol.; Dendrologia, 1 vol. 13 vols. folio, with numerous woodcuts. Bononiæ, 1599—1668.

2. *Historiam Naturalem* in Gymnasio Bononiensi

* I have much pleasure in thus acknowledging (by his initials) the assistance I have derived in these notices, from Mr. James Duncan, of Edinburgh; a zealous and able naturalist, already favourably known by his other biographical sketches, prefixed to the volumes of THE NATURALIST'S LIBRARY.

profitentis Ornithologiæ; hoc est, de Avibus Historiæ
Libri XII.   Bononiæ, 1646.

AMOREUX, N.—*Entomology.*

A physician at Montpellier.

Notice des Insectes de la France, reputés veni-
meux.   Paris, 1789. folio, with plates.

ANDERSON, JOHN.

A merchant and burgomaster of Hamburgh, born in
1674; died in 1743.

Histoire Naturelle de l'Islande du Greenland, &c.
Paris, 1750. 2 vols. 8vo.   Of this work, which we
have never seen, Cuvier remarks, that, " although
antiquated and superficial, it is still the principal
source of our information relative to the *Cetacea*."
We believe, however, that Dr. Scoresby's more recent
account of these regions, and of the *Cetacea,* is far
more to be depended upon.

ANONYMOUS. (CHR. FR. v. W**.)

Allgemeine historisch-physiologische Naturgeschichte
der Gervächse den Leibhabern, &c.   Gotha, 1791.
1 vol. 8vo. with 36 coloured plates.

ARGENVILLE, ANTOINE JOSEPH DESALLIERS D'.—
Conchology.

Comptroller of accounts at Paris.   Born in Paris,
1680; died in 1765.

L'Histoire Naturelle éclaircie dans une de ses
principales Parties, la Conchyliologie.   Paris, 1742.
A second edition, " augmentée de la Zoomorphose,"

*H 7

was printed in 1757. A third, " augmentée par M. M. Favanne," appeared in 2 vols. 4to.   Paris, 1780.

ARISTOTLE.

Every chronological record of the progress of natural history must commence with Aristotle, for he not only concentred in himself all the scattered and imperfect knowledge of his predecessors and cotemporaries, but added thereto the fruits of his own unrivalled genius, first giving to the .pursuit a character and consistency entitling it to the name of a science. Of the most comprehensive powers, his faculties seemed equal to every thing falling within the range of human contemplation; and the influence which his doctrines and opinions, once considered sacred as the responses of an oracle, have exercised on the general mind, has scarcely yet, after the lapse of so many centuries, and when the current of thought is flowing in such different channels, altogether ceased to be felt.

He was a native of Stagira, born in the first year of the 99th Olympiad — that is, about 384 years before the Christian era. His parentage, at least on the father's side, was highly respectable; his father, Nicomachus, being physician to Amyntas, king of Macedonia, father of Philip. Having lost both his parents at an early age, he was placed under the care of Proxenus and his wife, at Atarna, in Mysia. Attracted by the celebrity of Plato, he became a pupil of the Academy at the age of seventeen; and notwithstanding many alleged irregularities, he gained the unqualified approbation of that sage, by his astonishing industry and the extraordinary capacity of his mind. Plato used to call him " the soul of his school." On the death of his master (338 B. C.), he retired to Mysia; and three years after married Pythias, the adopted daughter of Hermeias, governor of certain cities in that country. The lady, however, died soon after, leaving an only daughter.

Not long after this, Aristotle was invited by Philip to the court of Macedon, to superintend the education of his son Alexander. Thither he accordingly went, and had the charge of his illustrious pupil for nearly eight years, enjoying the unbounded confidence of Philip, and the affectionate admiration of his son, a feeling which the latter ever retained, and took many opportunities of making known. When Alexander ascended the throne of Macedon, and began his extensive career of conquest, Aristotle went to Athens, and became a teacher of philosophy in the Lyceum. From his practice of instructing his disciples while walking in the shady avenues of the temple of the Lycian Apollo, which was within the precincts of the Lyceum, his scholars, as well as philosophy, obtained the name of Peripatetic ; and the sect, as is well known, became one of the most celebrated of ancient times. So much distinction could not then be enjoyed at Athens, without exciting the jealousy of rival philosophers ; and theological opinions of too refined a nature to suit the grossness of paganism, could not be taught without rousing the enmity of the hierophants : from these and other causes, Aristotle had many enemies, and they continued gradually to increase both in numbers and virulence. During the life of Alexander, however, they did not venture to take any overt step against him ; for although a considerable coolness now existed between Alexander and the philosopher, the latter still derived a degree of security from the terror inspired by the conqueror's name. That restraint having been removed by death, a conspiracy against his life was formed, and a charge of impiety preferred. Aware of the impossibility of obtaining justice from a partial and prejudiced tribunal, and having the fate of Socrates before his eyes, he secretly retired from the city, and took up his abode at Chalcis, in Euboea, where he soon after died, in the 63d year of his age (322 B. C.).

By his will (which happens to be still extant), Antipater of Macedonia was appointed his executor. His second wife, Herpylis (by whom he had an only son, named Nicomachus), is spoken of with much regard: the bulk of his fortune is left to Pythias, his daughter by his first wife, and Nicomachus; his library and writings to Theophrastus. His daughter, when of age, was to be offered in marriage to Nicanor, son of Proxenus; and failing him, to Theophrastus, his favourite scholar. The bones of his first wife were to be disinterred, as she herself had directed, and laid beside his own.

Into the *mare magnum* of the Aristotelian philosophy, a subject on which whole libraries have been written, it is not our province to enter in this place; and the pretty full account already given* of the περι Ζωῶν Ἱστοριας, which contains all that has come down to us of the Stagyrite's views on natural history, absolves us from the necessity of making any further remarks on that famous work.— J. D.

ARTEDI, P. — *Ichthyology.*

Celebrated as the father of systematic ichthyology. The intimate friend and disciple of Linnæus, who arranged and published his papers.

Pierre Artedi was born in 1705, in the Swedish province of Angermanland. As has happened to be the case with so many of those who ultimately became physicians and eminent naturalists, his early studies were conducted with a view to the church; but when he went to the university of Upsal, his attention was directed to medicine. It was here that he formed an intimate friendship with Linnæus, who was nearly of the same age, and engaged in similar pursuits. They studied together, made exploratory excursions into the country in company, and mutually aided each other by every means in their power. "He excelled me," says Linnæus, speaking of Artedi when a student at Upsal, "in chemistry, and I outdid him in the knowledge of birds and insects,

* Preliminary Discourse, p. 6.

and in botany." Their first separation was when
Linnæus set out on his Lapland journey; Artedi at the
same time departing for England, in furtherance of his
professional views. Before separating, the enthusiastic
young men made an agreement, that if either of them
should die, the survivor was to obtain possession of his
manuscripts and collections in natural history. They
both lived, however, to meet again; and this took place
at Leyden, in 1735, where they attended the prelections
of Boerhaave in company. Artedi had gone to Holland
for the purpose of obtaining his degree, but his extreme
poverty compelled him to postpone that measure. In
these circumstances, Boerhaave provided for him somewhat in the same way as he had previously done for
Linnæus: he recommended him to an apothecary of
the name of Seba, in Amsterdam, who had expended a
large sum of money in collecting a museum. Of this
he was then publishing a description, illustrated with
plates, and Artedi was employed to assist him in the
work. The branch of natural history to which Artedi
had latterly most attached himself, was ichthyology.
He had for some years been eagerly collecting materials
and making observations on this subject; and his manuscripts had now accumulated to such a degree, that he
contemplated the publication of them. His residence
with Seba afforded him further facilities for carrying
out this design, as his collection contained many fishes
which Artedi had not enjoyed any previous opportunity
of describing. But all his purposes were suddenly
frustrated, and the benefits which his industry and
abilities promised to confer on science prevented, by a
fatal accident which befel him in the thirtieth year of his
age. In a dark night he fell into a canal in the neighbourhood of Seba's house, and was drowned.

His manuscripts were fortunately still available for
science, but they were seized for debts. At the instigation
of Linnæus, his patron Clifford satisfied the creditors,
and the manuscripts were arranged and published by the
author's former friend and companion. The work ap-

I

peared in 1738, nearly three years after Artedi's death, with a memoir of the author from the pen of the editor. The work was a monument worthy of being raised by Linnæus in honour of his deceased friend. It was the most valuable and complete treatise that had appeared on the subject, and was confessedly of the greatest service to Linnæus, in his future arrangement of the class of animals to which it refers; and may be consulted with advantage even by the ichthyologist of the present day. The author fell into the old error of including the *Cetacea* among the true fishes. The work is entitled " Petri Artedi, Sueci Medici, Ichthyologia, sive Opera omnia de Piscibus;" divided into five parts;—1st, Bibliotheca Ichthyologica; 2d, Philosophia Ichthyologica; 3d, Descriptions of Genera; 4th, Synonyms of Species; 5th, Descriptions of Species. A corrected and enlarged edition, by Walbaum, appeared in Gripswald in 1788, 4to.; and the part relating to synonyms was republished by Schneider, with plates, at Leipsic, in 1789.

Artedi was likewise a very skilful botanist, and at one period of his life devoted much attention to the *Umbelliferæ*, of which he attempted to improve the arrangement by characters derived from the partial and general umbels. To commemerate his labours in this department, Linnæus named a genus of umbelliferous plants after him, *Artedia*.—J. D.

1. *Ichthyologia*, sive Opera omnia de Piscibus. Lugd. Bat. 1738. 8vo. pp. 102. and Index.

2. *Synonymia Piscium*, Græca et Latina, emendata, aucta, atque illustrata; sive Historia Piscium Naturalis et Literaria, &c. Auctore J. G. Schneider. Lipsiæ, 1789. 1 vol. 4to. with 3 plates. Cuvier alludes to another edition, edited by Walbaum, entitled *Artedi Renovatus*, published at Gripswald, 1788—89, 5 vols. 8vo. considerably augmented, but by an injudicious compiler. The following, also, by the same editor, is mentioned in Bohn's Catalogue.

3. *Genera Piscium*, emendata et aucta a J. J. Walbaum. Gripswald, 1799. 1 vol. 4to.

ASCANIUS, P.—*Zoology.*

Professor of Natural History at Copenhagen.

Icones Rerum Naturalium, or Figures enluminées d'Histoire Naturelle, 5 numbers, in oblong folio, each with 10 coloured plates, published between 1767 and 1779. The plates are badly executed, and mostly represent common objects; neither can the descriptions be commended.

AUDEBERT, JEAN BAPTISTE.—*Zoological Painter.*

An eminent zoological artist of France, born at Rochefort in 1759, died in 1800. Audebert must be ranked among the first scientific animal painters of his age. There is a life and animation in his designs, which show an intimate acquaintance with nature; his attitudes, in general, are easy and graceful, at the same time they are well calculated to display the peculiarities of each subject to advantage. Yet, in his delineation of birds, he completely failed: he gave a beautiful and correct representation of a stuffed specimen, but he could neither throw life nor action into its form; but this need not excite surprise, for no zoological painter has yet excelled in both departments. His work on quadrupeds deserves a place in every library intended to show whatever is most perfect of its kind. Most of the copies are printed in colours,—an art peculiarly adapted for quadrupeds.

Histoire Naturelle des Singes et des Makis; folio, avec 63 planches dessinées d'après les individus empaillés du Muséum. Paris, 1800.

AUDOUIN, JEAN VICTOR, M.D.—*Entomology.*

We hesitate not to assign to this name the reputation of the most philosophic, profound, and accurate entomologist now in Europe. M. Audouin holds the honourable office of sub-librarian to the French Institute, and was formerly the chief assistant, in that establishment, to Lamarck and Latreille.   He has published numerous essays of great value in the *Annales des Sciences Naturelles*, both singly and conjointly with his friend, Dr. Milne Edwards : most of these, unfortunately, we do not possess ; nor are they very accessible to the English student.   M. Audouin has likewise co-operated with MM. Le Chat and Geoffroy St. Hilaire ; but the only separate publication of his, in conjunction with Dr. Edwards, is the following :—

Résumé d'Entomologie, ou d'Histoire Naturelle des Animaux Articulés.   Paris, 1829. 2 vols. 18mo.

His Natural History of the Marine Animals of France, long expected by the scientific world, and on which he laboured so assiduously with his friend Dr. Edwards some years ago, has only just commenced being published.   M. Audouin was born in Paris, in April, 1797.

AUDUBON, J. J.— *Animal Painter.*

Celebrated for being the first ornithological painter of the age, and for having brought to its completion the largest sized collection of plates of birds ever published.

1. *The Birds of America.* 5 vols. 8vo. Atlas folio.
2. *Ornithological Biography.*
3. *Synopsis* of the Birds of North America. 1 vol. 8vo. London, 1839.*

Of the first of these publications we have already spoken in terms of unqualified praise, and we rejoice it has been brought to a completion, although its enormous expense precludes us from possessing, and consequently of consulting, it.   We cannot, however, speak in equal terms of approval of the biography, which is, in fact, the text of the plates.   Mr. Audubon is confessedly only a field naturalist, not a scientific one.†   He can

* The author confesses his obligations to Mr. Gilvray.
† It is singular how two minds, possessing the same tastes, can be so diversified, as to differ *in toto* respecting the very same objects.   During the whole time of Mr. Audubon's residence in Paris, he only visited the Ornithological Gallery twice, (where I was studying for hours, almost daily,

shoot a bird, preserve it, and make it live again, as it were, upon canvass; but he cannot describe it in scientific, and therefore in perfectly intelligible, terms. Hence he found it necessary, in this part of his work, to call in the aid of others; but being jealous that any other name should appear on the title page than his own, he was content with the assistance of some one who, very good-naturedly, would fall in with his humour. Thus, in a scientific point of view, the characters of the two publications are very different. From the same cause, also, we must attribute the frequent introduction of young birds, as new species discovered by himself. A want of precision in his descriptions, and a general ignorance of modern ornithology, sadly disappoint the scientific reader; all which are discerned in the "Biography," and are very striking in the "Synopsis," where he rejects established names *, and coins new ones of his own. The letter-press, however, is relieved by a series of well-written episodes, illustrating the manners, the habits, and the scenery of North America and its inhabitants. We only suspect that Mr. Audubon participates in the almost universal blemish of his countrymen, in colouring his narrations (not his paintings) somewhat too highly. He has, we believe, returned to live upon his property in the United States; but one of his sons is settled in London as an artist, and will, no doubt, inherit something of his father's pictorial talents.

AZARA, DON FELIX DE. — *General Zoology.*

Pre-eminently distinguished as the Spanish naturalist, was born on the 18th of May, 1746, in the province of Aragon, at Barbunales, near Balbastro. He had an only and elder brother, Don Nicolas, who also rose to eminence, but in a very different sphere, viz. as a diplomatist and patron of the fine arts; and yet it is some-

---

for the purpose of calling upon me ; and even then he merely bestowed that sort of passing glance at the magnificent cases of birds, which a careless observer would do while sauntering in the room.

* As *Myiodoctes* for *Sylvicola.* Several other of M. Audubon's new genera, I am obliged to confess, are quite unintelligible to me.

what remarkable, that in some of our latest biographical
accounts, the two brothers have been confounded together, or, rather, we should say, compounded into one.
Felix commenced his studies at the university of Hues-
ca, and then removed to the Military Academy of Bar-
celona, but rarely, during the period, visiting his parental
roof. At the age of eighteen, he commenced his career
as a soldier; was soon appointed to the corps of engineers,
in which he rose to the rank of lieutenant-colonel; and
after service for nearly forty years, was appointed brigadier-general in the army. In the year 1775, when
holding the rank of lieutenant of engineers, forming part
of the force which at that time attacked Algiers, being
among the first who landed, he was wounded in the
side by a cannon ball, and for a time was left for dead.
Though, for years, this wound occasioned much trouble,
yet, in the long run, he completely outlived all its un-
pleasant effects.

It was in the year 1781, when doing duty as lieutenant-colonel at St. Sebastian, that, during the night,
our soldier received an order to set off instantly to Lisbon, and he repaired thither without a moment's delay.
He was immediately despatched across the Atlantic to
the Spanish settlements in South America; as it was
afterwards explained to him, for the purpose of acting
as one of the government commissioners in settling the
limits of the Spanish and Portuguese possessions in that
vast continent: and he was, at the same time, honoured
with the rank of captain in the Spanish navy. With
the most commendable zeal, he set to work in the ac-
complishment of his arduous task, which unfortunately,
however, did not depend upon his single exertions.
The co-operation of the Portuguese commissioners was
essential; and, as it speedily appeared that the execution
of the details of the treaty on which they were acting,
would deprive Portugal of a portion of her provinces,
every obstacle that could be found was thrown in the
way. Nor was this disposition confined to the commissioners. The viceroy, and not only the Portuguese,

but the Spanish likewise, did little or nothing to check
these delinquencies, and hence Azara was chagrined and
almost broken-hearted. Without leave, he could not
retire from his allotted post; and that permission he
could not obtain. It was under these circumstances
that the inherent activity of his mind in a great measure overcame the disadvantages of his situation; and
while considering himself in little better than a state of
banishment, past the meridian of a life spent in camp
and field, he speedily formed that plan of study and
exertion, on the execution of which rests the chief share
of that fair fame and celebrity for which he is distinguished. His previous education had, to a certain extent, prepared him for geographical investigations; and
he undertook many extensive travels over these vast
regions, equal in extent to nearly the whole of Europe,
with this grand object in view. Day and night he took
solar and sidereal observations, and, by innumerable abstruse astronomical investigations and calculations, accurately settled many a latitude. These long journeys
again brought him into frequent and close contact with
the aboriginal tribes, or nations as they are called, to
the number of between thirty and forty; and these, and
the mixed breeds, and those of pure Spanish descent, he
carefully studied and minutely described; and thus, beginning with man, the self-taught naturalist carried
down his observations throughout the whole animal
series, during a period of nearly twenty years. He
thus prepared a work on the quadrupeds of the country,
and a still more elaborate one on the birds, discussing
also the *Amphibia* and insects, and paying much attention to botany, to the productions of the country, its
minerals, climate, &c. &c. The materials of these works,
including his *Voyages* and an Atlas, were collected on
the spot, and with little farther help than a Spanish
edition of Buffon latterly supplied. On at length obtaining leave to return to Europe, in 1801, he arranged
these materials, and published them, to the high gratification of all naturalists.

In the year 1802, Don Felix rejoined his brother, at
that time filling the post of Spanish ambassador at Paris,
who was so delighted with his society, that he prevailed
with him to resign his rank of brigadier-general, and
take up with him a permanent abode.   This gratifying
enjoyment was, however, short-lived, Don Nicolas ex-
piring in his brother's arms, in January, 1803.   A conqueror's mad ambition soon afterwards compelled him
to return to his native land.   There he was speedily
appointed member of the board in which was centred
the home government of the Spanish Transatlantic affairs; and in 1805 he was called to Madrid, to serve
his country at a more urgent post.   When the French
edition of his work was published, in 1806, one of the
consequences of the war was, that he could not even
obtain a copy of it; and, after this date, we are not
aware that any information respecting him has reached
this country, — a blank which, we should think, might
now easily be supplied.

1. *Essai sur l'Histoire Naturelle* des Quadrupèdes
du Paraguay, traduit sur le Manuscrit par M. Mo-
reau de St. Méry. Paris, 1801. 2 vols. 8vo.   The original work is entitled *Appuntamentos* para la Historia
Naturel de los Quadrúpedos del Paraguay y Rio de la
Plata.   Madrid, 1702. 2 vols. 4to.

2. *Voyages dans l'Amérique Méridionale*, de 1781
jusqu'en 1801.   Traduits par M. Walkenaer.   Paris,
1809. 4 vols. 8vo. and 1 of 4to. plates.   The two
last are edited by Sonnini, and contain the ornithological portion ; unfortunately, only the Spanish vul-
gar names are given, so that the student remains ignorant of the systematic nomenclature.

BÆRII, NICHOL. — *Ornithology*.

Ornithophonia, sive Harmonia Melicarum Avium
juxta Naturas, Virtutes, et Proprietates suas Carmine
Latino-Germanico. Breinæ, 1695. 4to., with woodcuts. — (*Bohn's Cat.*)

BAJON. — *General Zoology.*

Chief surgeon to the colony of Cayenne.

Mémoires pour servir à l'Histoire de Cayenne. Paris, 1777. 2 vols. 8vo.   Containing several notices on the native animals.

BARBUT, JAMES.

Genera Insectorum of Linnæus.   London, 1781. 4to., with 22 coloured plates.
Genera Vermium, in two parts.   London, 1783 — 88.   Coloured plates.

BARRÈRE, PIERRE. — *Ornithology.*

Professor in the University of Perpignan ; died in 1755.

1. *Ornithologiæ* Specimen novum, sive Series Avium in Ruscinone, Pyrenæis Montibus, atque in Gallia Æquinoctiali observatarum.   Perpignan, 1745. 4to. pp. 84. pl. 1.
2. *Essai* sur l'Histoire Naturelle de la France Equinoxiale.   Paris, 1741. 12mo.

BARTON, B. SMITH. — *General Zoology.*

The father of natural history in America ; for many years, up to the period of his death, Professor of Botany in the University of Philadelphia.   He died at an advanced age, in 1816.

1. *Memoir* on the Fascination attributed to the Rattlesnake.   1 vol. 8vo.   Philadelphia, 1796.
2. *Facts,* Observations, and Conjectures, on the Generation of the Opossum.   8vo.   Phil. 1801.
3. *Sirens Lacertina,* Notices of, and of another Species of the same Genus.   8vo.   Phil. 1808.

4. *Memoir on a Reptile* called the Hellbender. 8vo. 1812. The *Salamandra gigantea* of subsequent authors.

BAUDET DE LA FAGE, MARIE JEAN.—*Entomology.*

Essai sur l'Entomologie du Département du Puy-de-Dôme. Monographie des Lamelli-antennes. Clermont, 1809. 8vo.

BASTER, JOB. — *Entomology.*

Born in 1711. Practised as a physician at Harlem, was elected an F.R.S. of London, and died in 1775.

Opuscula subseciva. Harlem, 1764—5. 2 vols. in 1. 4to., with figures. This contains remarks on the use of the antennæ of insects.

BEAUVOIS, PALISOT BARON DE. — *Entomology.*

1. *Mémoire* sur un nouveau genre d'Insectes (Atractocerus), in an 8vo pamphlet.

2. *Insectes récueillés* en Afrique et en Amérique. Paris, 1805. folio. The plates are printed in colours, and are very good.

BECHSTEIN, J. M.—*Ornithology.*

An eminent Saxon naturalist.   One of the first authorities on the birds of Europe.

Gemmeinnutzige Naturgeschichte Deutschlands, Zweyte auflage,—und Ornithologisches Tasschenbuch von und für Deutschland, or, the Natural History of the Quadrupeds and Birds of Germany. Leipsic, 1801—1809. 4 vols. 8vo.

Naturgeschichte der Stubenvögel. Gotha, 1795. 4 coloured plates, 8vo. Another edition was printed at the same place in 1812, and contains 16 plates.

BELANGER, CHARLES. — *Oriental Traveller.*

Zoologie du Voyage aux Indes Orientales, par le
Nord de l'Europe. Paris, 1824. 1 vol. 8vo. 40 plates
in 4to. coloured. Of this we have only seen three
numbers. The descriptions are by Geoffroy St. Hilaire,
Guérin, Lesson, &c. : many of the subjects, however,
supposed to be new, are not so.

BELL, THOMAS. — *Erpetology.*

A learned and eminent erpetologist, and Professor of
Zoology in King's College, London. Besides numerous
papers in the *Zoological Journal* and other scientific
Transactions, he has commenced a splendid work on
the tortoises.

　1. *Monograph of the Testudinata*, with Descriptions
in Latin and English. Folio, 5 coloured plates in
each.
　2. *British Reptiles*, A History of. 1 vol. 8vo., with
beautiful woodcuts. London, 1840.
　3. *British Quadrupeds*, A History of; uniform with
the last. London, 1840.

BELON, PIERRE. — *Ornithology.   Ichthyology.*

Pierre Belon was the most distinguished of a small
brotherhood of naturalists who flourished about the
middle of the 16th century, and who devoted themselves
chiefly to the investigation of the history of fishes. They
were the first who took up that subject in a philosophical
manner after the revival of learning, and they may be
said to have laid the foundations of modern ichthyology.
Although cotemporaries, they do not, however, seem to
have had any intercourse with each other, and each of
their works is therefore to be regarded as the exclusive
result of personal observation and study. The individual
to whom the present notice refers, was born at Souletière,

a hamlet in the parish of Oisé, in Le Maine, about the
year 1518. We possess no detailed account of his family and descent — both, probably, being very obscure,
and there is the same want of information regarding his
boyhood and education. When he became of age to
think of a profession, that of medicine was chosen ; and
botany, as a requisite attainment, was attended to, and
soon became, as it has so often done in similar cases, his
favourite pursuit. It appears that he was indebted for
his education to René du Bellay, bishop of Mans, William Duprat, bishop of Clermont, and some other dig-
nitaries of the church. Through their influence, he was
placed under the tuition of Valerius Cordus, Professor of
Natural History at Wirtemberg ; and having gained the
favour of his teacher, he was selected to accompany him
into Germany and Bohemia, in some excursions undertaken with a view of examining the natural history of
these countries. On returning from one of these expeditions, he was arrested, it does not appear for what cause,
at Thionville, and might have been detained for a considerable time, but for the interference of a chivalrous
gentleman, who obtained his release because he was a
co-patriot of a favourite poet of his.

The next important incident in his life was his journey
through Greece, Egypt, Palestine, and Asia Minor, during
which he collected many interesting objects in natural
history, and likewise in relation to the state of the arts,
agriculture, &c. in these countries. These he arranged
while residing in Paris, in 1550, and published several
works, both on natural history and subjects of general
interest to a traveller in the countries he had visited. He
set out on another journey in 1557, and traversed Italy,
Savoy, Dauphiny, and Auvergne. He had now acquired
distinction for his learning and powers of observation ;
and on his return, Charles IX. assigned him a residence
in the Château de Madrid. Here he occupied himself
with the preparation of a work on agriculture, in translating Dioscorides and Theophrastus, and in other
literary and scientific matters. But his career was sud-

denly terminated by the hands of an assassin, who murdered him in the Bois du Bologne, as he was on his way to Paris. This took place in 1564, when he was in his forty-sixth year.

Besides his *Travels*, he published various works on ichthyology, botany, and ornithology. His book on the last of these subjects is frequently cited by Buffon, and was considered of considerable utility at the time of its appearance. His *De Arboribus Coniferis*, &c. was published at Paris in 1553. An essay, entitled " Remonstrances on the Neglect of introducing and cultivating Foreign Plants," is said to have led to the formation of a kind of botanical garden at Montpellier, even before one existed in the capital. The plants and animals which he observed in Arabia and Egypt were figured in a work which appeared in 1557. In acknowledgment of his services to botany, Plumier has dedicated an American genus of plants to his memory by the name of *Belonia*. But it is for his contribution to ichthyology that he deserves most to be held in honour. His principal production on this subject appeared in 1551, in 4to. ; and it was succeeded, as will be seen by a list of his works appended to this notice, by several others of a like nature. Many of the species are figured from wood engravings.

A singular charge of plagiarism has been brought against Belon, which would have been unworthy of notice, had it not found its way into one or two works of authority. It is alleged, that he accompanied Pierre Gillius, an individual of some distinction, in a journey to Rome, in the capacity of attendant ; and that when Gillius died, an event which took place at Rome in 1555, Belon appropriated his manuscripts, and afterwards published them in his own name. It is enough to show the improbability of such an occurrence, to consider that Belon had already published some of his best works, and that there is nothing in any of those that followed which he was not competent to produce; the singularity of style, and mode of treating, as well as the nature of the subjects themselves, all indicate them to be the production of the same hand. As a writer, Belon is noted for the simplicity and force of his style.

1. *Etranges Poissons Marins*, Histoire Naturelle des, et Description du Dauphin, &c. 1551. 4to.

2. *Poissons*, Histoire des. Transv. 1551. 8vo.

3. *Oiseaux*, Histoire Naturelle des. 1553. folio.

4. *Observations* faites dans ses Voyages en Orient. 1553. 4to.

BENNETT, ED. TURNER. — *General Zoology.*

A surgeon, settled in London. He was one of the early and most active promoters of the Zoological Society, of which he was secretary until the time of his death (1838). Although the author of many valuable papers, he published no separate work, beyond editing an account of the Zoological Gardens.

BENNETT, J. WITCHURCH. — *Ichthyology.*

Fishes of Ceylon; a Selection of the most remarkable and beautiful Fishes found on the Coast of Ceylon. 6 parts, 4to. 30 coloured plates. London, 1830. This work is valuable, on account of the figures and descriptions having been taken from fresh specimens : but it is not very scientific.

BERGEN, C. A. DE. — *Conchology.*

Classis Conchyliorum. Norimb. 1760. 1 vol. 4to.

BESEKE, J. M. T. — *Ornithology.*

Materials for the History of the Courland (in German). Mitfau et Leipz. 1792. 1 vol. 8vo.

BESLER, M. R.

Physician at Nuremberg. Born in 1607 ; died in 1661. *Rariora* Musci Besleriani.* 1716. folio.

---

* Percheron gives the title of this work as " Gazophilacium Rerum Naturalium ;" and mentions two other editions, one in 1642, the other in 1733.

BEWICK, THOMAS. — *General Zoology.*

The reviver of wood engraving; and the most eminent artist, in that department, of his age.

1. *History of British Birds.* The Figures engraved on Wood, by T. Bewick. Newcastle, 1805. 2 vols. royal 8vo. Another edition was published in 1822.

2. *Figures of British Land Birds,* engraved on Wood, by T. Bewick ; to which are added a few foreign birds. Newcastle, 1800. 1 vol. 8vo. 133 plates. To these plates there is no text. The British portion is printed from the same blocks as those used for the former work. The figures of foreign birds (14 in number), not having been taken from living specimens, are inferior to the others : they were intended for a general work on birds, a design afterwards abandoned.

BILBURG, GUST. JOHN. — *Entomology.*

1. *Monographia Myladridum.* Holmiæ, 1813. 8vo. with 7 plates.

2. *Enumeratio* Insectorum in Museo Auctoris. Holmiæ, 1820. 4to. besides other papers in the Upsal Transactions, &c.

BLAINVILLE, HENRI DUCROTAY DE. — *Zoology.*

Professor in the French Museum and Garden of Plants. A skilful anatomist, who has proposed great reforms in systems and nomenclature.

De l'Organisation des Animaux, ou Principes d'Anatomie comparée. 4 vols. 8vo. Paris.

BLOCH, M. E. — *Ichthyology.*

Mark Eliazar Bloch was of Jewish descent, and born at Anspach in 1723. His parents were in somewhat indigent circumstances, and his early education was in consequence almost entirely neglected. We are told, that

even up to the age of nineteen, he had no acquaintance
with the learned languages or the literature of his times,
his reading having been confined to a few of the Rab-
binical works of his tribe. A desire for learning was
first excited, when he went to reside in the family of a
Jewish physician in Hamburgh ; and once having entered on the path to knowledge, he followed it eagerly.
He acquired some acquaintance with German, and a poor
catholic priest taught him Latin ; while some instructions in anatomy and other branches of medical learning
were communicated to him by the physician with whom
he resided. He then went to Berlin, where several of
his relations lived ; and continued his studies, both in
natural history and medicine, with such diligence and
ability, that his progress was unusually rapid.

He was thus enabled to obtain his degree of M.D.,
which was granted him at Frankfort on the Oder. He
then returned to Berlin, and commenced practice as a
physician ; and here he resided for the remainder of his
life. His death took place on the 6th of August, 1799.

As a naturalist, he enjoyed the patronage of Martini.
He by no means confined his attention to one branch
of the subject, but he laboured most in ichthyology,
and it is for what he accomplished in this department
that the gratitude of succeeding inquirers is due. His
principal work is a " Natural History of Fishes, particularly those of the Prussian States," in four parts,
4to. Berlin, 1781—2. A work on foreign fishes, and
another on the fishes of Germany, subsequently appeared.   All these were afterwards remodelled and
published as one work, entitled " Ichthyologia, or General and Particular History of Fishes." Berlin, 1785,
in 12 vols. 4to.   This was reprinted in 1795, with the
text translated into French, and illustrated with 432
coloured plates.   The expense of these plates was defrayed by the various individuals to whom they are
respectively dedicated, and who justly looked upon the
work as of national importance. — J. D.

1. *Ichthyologie*, ou Histoire Naturelle des Poissons. Berlin, 1785—97. 12 parts, folio. Cuvier mentions the number of plates as 452, Wood as only 429. The first six parts comprise the fishes of the northern hemisphere, accurately described and tolerably well figured: the latter six parts are very rare, the unsold copies having been destroyed by a fire.

2. *Another edition*, in 10 vols. 18mo., was published at Paris in 1801, with 157 plates.

2. *Systema Ichthyologiæ*, edited by Schneider. Berlin, 1801. 2 vols. 8vo. pl. 110. This, which we have never seen, includes many species taken from other authors, "mais sous une méthode bizarre " (*Cuv.*).

3. *Traité sur la Génération* des Vers Intestines (in German). Berlin, 1782. 4to.

BLUMENBACH, JEAN FRED. — *Comparative Anatomy.*

A celebrated anatomist, and Professor of Medicine and Natural History at Gottingen.

1. *Manuel* d'Histoire Naturelle. Metz, 1804. 1 vol. 8vo. This elaborate work has gone through numerous editions in Germany: the French translation here mentioned, is that cited by Cuvier, and is made from the eighth German edition.

2. *Figures* of Natural History (Abbildungen). Gott. 1796—1810. Ten numbers, each with 10 plates.

BOCCONE, PAULO. — *Zoology and Botany.*

Paul Sylvius Boccone deserves to be held in honourable remembrance for the efforts he made in favour of natural history, at a time when very few thought the subject worthy of attention. He was of a noble family, and born at Palermo in the year 1633. Botany was the subject he chiefly cultivated, although he did not altogether neglect other departments. He travelled

K

through the greater part of Europe, to make himself acquainted with the naturalists of the day, and obtain by
intercourse with them some of the knowledge which
they had acquired. By the advice of the abbé Bour-
delot, whose friendship he had gained at Paris, he published a work entitled " Researches and Observations in
Natural History." Amsterdam, 1674. In the course of
his peregrinations he visited London, and became acquainted with Hatton, Sherard, Monson, &c. While in
England, he drew up the work entitled " Icones et Descriptiones rariarum Plantarum Siciliæ, Melitæ, Galliæ,
et Italiæ, &c.," which was printed at Oxford under the
care of Morison, in 1674. It is a quarto volume, containing 52 plates. Boccone afterwards went to reside
at Venice, where he published a second botanical work
very similar in its design to the former. Another of a
more miscellaneous character afterwards appeared, containing, among other matters, remarks on corals and
the eruptions of Mount Etna. He had the honour of
being nominated botanist to the grand duke of Tuscany,
a prince who did more for the promotion of science than
most others of that period. When about the fiftieth
year of his age, Boccone became dissatisfied with the
world, and retired to a monastery, assuming at the same
time the name of Sylvius. It was in a monastery near Palermo that he died, on the 22d of December, 1704, aged
seventy-one. He is said to have left many manuscripts,
particularly a " Natural History of Malta," which were
never made public. *Bocconia* is a genus of papaver-
aceous plants, so named by Plumier in honour of this
naturalist. — J. D.

BODDAERT, PIERRE. — *Mammalogy.*

Physician and senator of Flessingue, in Zealand.

Elenchus Animalium. Vol. I. Quadrupedes. Rotterdam, 1785. 8vo. pp. 174. This work was never
continued.

BOHATSCH, J. B. — *Malacology.*

Professor at Prague : died in 1772.

De quibusdam Animalibus Marinis, &c. Dresden, 1761. 4to. A work containing many good observations on several mollusks and zoophytes.

BOÏE. — *General Zoology.*

A young naturalist, native of Keil, of great promise, who was sent to explore the zoology of Java, where he unfortunately fell a victim to the climate. He seems to have published many papers in the *Isis,* but we are unacquainted with any separate publication bearing his name.

BOISDUVAL, J. A. — *Entomology.*

A nominal physician, and late curator of the cabinet of count Dejean.

1. *Essai* sur un Monographie des Zyganides. 1 vol. 8vo. Paris, 1829. To which is added, Europæorum Lepidopterorum Index Methodicus.

2. *Hist. Générale* et Iconographie des Lépidoptères de l'Amérique Septentrionale, conjointly with Le Conte, in 8vo. numbers. (See LE CONTE.)

3. *Iconographie* et Hist. Nat. de Coléoptères d'Europe. 8vo. Paris, 1827. In periodical numbers, conjointly with count Dejean ; besides various papers in scientific periodicals, and a poor continuation of Buffon, on the *Lepidoptera.*

BOJANUS, LOUIS HENRY. — *Erpetology.*

An excellent German erpetologist, who became one of the professors at Vilna, where he died in 1828. Besides several memoirs in the *Isis,* he wrote

A Monograph of the Freshwater Tortoises of Europe. Vilna, 1819. 1 vol. folio.

BONAPARTE, CH. LUCIEN, PRINCE OF MUSIGNANO.* —
General Zoology.

This illustrious and learned zoologist is the eldest son
of Lucien Bonaparte, prince of Canino. He remained for
many years in the United States, but has since resided
in Rome. He is connected by marriage with lord Dudley
Stuart, and has made frequent journeys to England. All
his works are valuable, and evince considerable acuteness
and research. Besides his numerous papers in the American and other Scientific Transactions, he is chiefly
known as the author of the following: —

1. *American Ornithology;* being a Supplement to
that of Wilson, and uniform with the quarto edition.
Of this we have seen three volumes.

2. *Iconographia* della Fauna Italica. 4to. with
most admirable plates, complete in 24 parts. Rome,
1832—39. Not being possessed of this work, we
could not cite it for the fishes of the Mediterranean,
mentioned in our ichthyological volumes.

BONNANI, PHILIP. — *Malacology.*

A professor in the Jesuits' College at Rome. Born
in 1638; died in 1725.

Recreatio Mentis et Oculi in Observatione Animalium Testaceorum. Romæ, 1684. small 4to.

BONNATERRE, ABBÉ. — *Erpetology. Ichthyology.*

Professor of Natural History at Tulle. He is only
known as an author, from having written the account of
the reptiles and fish in the *Encyclopédie Méthodique.*
He likewise, as Cuvier observes, " superintended the
engraving of the vertebrated animals" for the same work;
of which it would be impossible, perhaps, to name a

---

* The death of his father, just announced in the papers, will probably
elevate him to the title of Prince of Canino.

worse collection, excepting those in some of the editions of Buffon, from which many are copied.

BONNELLI, FRANÇOIS. — *Ornithology. Entomology.*

Director of the Museum, and Professor of Zoology at Turin. An acute and indefatigable entomologist.

1. *Catalogue des Oiseaux* du Piémont. 1811. 4to.
2. *Observations Entomologiques.* These two valuable essays, in which a luminous account is given of the Linnæan genus *Carabus,* have, we believe, been printed separately; although originally they were inserted in the Memoirs of the Academy of Sciences at Turin.

BONNET, CHARLES. — *Entomology.*

A celebrated philosopher and philosophic naturalist. Born at Geneva in 1720; died in 1793. As a zoologist he is best known by his

Traité d'Insectologie. Paris, 1745. 2 vols. 8vo.

BONTIUS, I. — *Zoology and Botany.*

A learned physician of the seventeenth century, who was at the head of the Dutch medical establishment in Batavia. His work was published in conjunction with that of Piso and Marcgrave. It is very curious, as being one of the first on the natural history of India.

Historiæ Naturalis et Medicæ Indiæ Orientalis Libri VI. There is an English translation, which we have not seen, thus entitled: — I. Bontius's Account of the Diseases, Natural History, and Medicines of the East Indies. London, 1769. 1 vol. 8vo.

BORHKAUSEN. — *Entomology.*

Naturgeschichte der Europäischen Schmetterlinge nach Systematischer ordnung. Frankford, 1788—94. 5 vols. 8vo.

BORLASE, WILLIAM. — *Natural History, &c.*

A clergyman, who investigated the history and natural productions of his own county, — a work still of real value. He was born in 1696, and died in 1772.

The Natural History of Cornwall. Oxford, 1758. folio. A second volume, it has been said, was subsequently published, which has now become so very rare that we have never seen it.

BORN, IGNATIUS BARON. — *Conchology.*

A celebrated mineralogist in the service of Prussia. Born in 1742; died in 1791.

Testacea Musei Cæsarei Vindobonensis, disposuit et descripsit. Vind. 1780. folio, pl. 18. This, the only zoological work of the author, is principally valuable for the excellency of its figures.

BORY SAINT VINCENT. — *Zoology and Botany.*

The most eminent of all the naturalists who accompanied the unfortunate expedition of captain Baudin as far as the Isle of France. He is better known as a botanist than as a zoologist; but in both his superior talents are conspicuous. His *Voyage aux Quatre principales Isles d'Afrique* we have not seen, nor are we aware that he has published any separate zoological work; but many of his papers in the scientific journals of France relate to animals, and all which proceeds from his pen is valuable. He holds, we believe, the rank of colonel in the French service.

BOSC, LOUIS. — *Malacology.*

Member of the French Academy of Sciences. An experienced and zealous naturalist, who has written many papers in the French journals.

Histoire Naturelle des Vers, &c., in the small 12mo. or Deterville's edition of Buffon.

BOSMAN, WILLIAM. — *Traveller.*

Voyage en Guinée. Utrecht, 1705. 1 vol. 8vo. In this volume, Cuvier observes, will be found many interesting observations on the native animals.

BOURGUET, LOUIS.

Professor at Neufchâtel. Born in 1678; died in 1642.
Traité des Pétrifications. Paris, 1742. 4to.

BOWDICH, TH. ED. — *General Zoology.*

Of this enthusiastic traveller and naturalist, we have been favoured with some private information, which forms the ground-work of the following notice : —
The family of Bowdykes, in Dorsetshire, who trace their lineage up to the Saxons, were the ancestors of Thomas Edward Bowdich. He was born at Bristol, in 1790; at the grammar-school of which city he first commenced his education : as his proficiency was remarkable, he was soon removed to Corsham, in Wiltshire; where he soon distinguished himself, even among a number of competitors. He was entered at Oxford, and intended for one of the liberal professions ; but his father, who appears to have been somewhat fickle, changed his intentions towards him. Uncomfortable and disappointed at home, he was probably induced to turn his thoughts to Africa, where his uncle was then second in command of the English posts on the Gold Coast. Accordingly, in 1814, he embarked for that country; and was soon followed by his wife. Some unexplained circumstances, however, induced him to return again to England the same year ; when, an expedition to Ashantee having been resolved on, Bowdich

was formally appointed to its chief command, under the
title of conductor. But a cruel disappointment awaited
him. No sooner had he returned again to Africa, — naturally elated, as every ardent mind would be, — than
Mr. Hope Smith, the governor of Cape Coast, took upon
himself to thwart the authorities in England; he at once
informed Mr. Bowdich, that, in his opinion, he was too
young to conduct a mission of such importance; and
therefore he, the governor, could only place him as
second in command. Mortifying as this conduct would
have been to any man, it was still more keenly felt
by one of an ardent and sanguine temperament: yet
this did not alter his determination of proceeding in
what he had begun. The event showed the weak judgment of the governor, and the energy of him whom he
had taken upon himself to displace. The embassy
marched to Coomassee in April, 1813. The appointed
leader, unfit for his charge, was soon recalled; and
every thing was placed under the orders of Bowdich.
His prudence and energy were now fully developed.
Having perfectly succeeded in the difficult and dangerous task allotted to him, he once more returned to
England, in 1817. The publication of his travels, which
immediately followed, at once established his reputation; and he became known throughout Europe, not
merely as a successful African traveller, but as a man
of tried courage and decided talent. Can we add, that
these qualities were appreciated by the small knot of
individuals who more immediately directed this undertaking? Far from it. Having now proved his
own strength, and practically seen what *could* be done,
both for science and civilisation, in that neglected country, he was most anxious to return there, and carry on
his operations on a larger scale. But he neither met
with reward for the past, nor encouragement for the
future. Such is the usual fate of genius; and such the
wretched consequences of placing men in authority over
things they understand not, merely because they possess parliamentary interest. No wonder, therefore,

that the high-minded spirit of Bowdich turned indig-
nantly from the authors of such injustice. He quitted
the thankless service of the government, and determined to do what he could in the cause of Africa by
his own unassisted resources. To increase his means,
he went with his accomplished wife to Paris, where he
was enthusiastically received by Cuvier, and all those
whose talents reflected honour upon whomsoever they
admitted to their friendship. In such society, — so different from the form, the ceremony, and the display of
that in London, — did Bowdich pass near three years
and a half; supporting himself and his family by his
pen, and daily becoming more and more qualified for the
enterprise he had in hand. Here, also, he wrote those
useful compendiums subsequently noticed, as text books
for future travellers, on the Cuvierian arrangement of
animals. The French government, who always see and
do these things better than our own, used every means
in their power to engage the service of Bowdich as
traveller under their authority : but this he declined,
from a feeling — to us, a fastidious one, after the treatment he had received — of not devoting his services
to any other country than his own. His preparations were almost completed to depart for Africa,
when he received an offer from sir Charles M‘Carthy,
then in Paris, to proceed to Sierra Leone, where they
were to combine operations; and this offer, it appears,
he accepted. Intending to take Lisbon in his way, he
sailed for that capital in 1822, where he collected materials for a narrative of the Portuguese discoveries in
Africa: he then proceeded to Madeira ; where he had
the mortification of being kept for more than twelve
months, waiting for a conveyance onward. Unable to
meet with a vessel sailing direct for Sierra Leone, he
went to the Cape de Verde islands, and thence to the
river Gambia. Here, again, he was obliged to halt; yet
his active mind turned every thing to advantage : he
employed himself in making trignometrical surveys, and
in collecting every information, of this part of Africa.

Regardless of his health, he frequently exposed himself to the sun while thus occupied : a slight fever was the result ; and this he increased by going out one night to observe the satellites of Jupiter. These imprudences unfortunately proved fatal ; for, after a fortnight's illness, he expired in the arms of his wife, on the 11th of January, 1821.

His accomplished widow (now Mrs. Lee) is well known in the literary world, as an elegant and beautiful writer of fiction, no less than of reality.

1. *Ornithology* of Cuvier, An Introduction to the, for the Use of Students and Travellers. Paris, 1821. 8vo. pp. 86. pl. 21. A very useful compilation of Cuvier's system ; it also contains figures of those parts which constitute the leading generic characters.

2. *Conchology*, an Introduction to ; including the Fossil Genera. Paris, 1822. 2 parts, 8vo. with numerous figures. The *Quadrupeds* were treated of in a similar manner.

BOYS, WILLIAM. — *Conchology.*

An assiduous investigator of British shells.

Testacea minuta rariora nuperrime detecta in Arenâ Littoris Sandvicensis a Gul. Boys, multa addidit et omnium figuras delineavit G. Walker. Lond. 4to. pp. 25. pl. 3.

This work, observes Mr. Wood, is erroneously quoted, by many writers, as being the production of Walker, who, in fact, was only the artist employed in the execution of the plates.

BRANDER, GUSTAVUS. — *Conchology.*

An elegant scholar, and a gentleman of fortune, who formerly possessed the estate (High Cliff, Hants) upon which are situated the Cliffs of Hordwell and Barton, so celebrated for their innumerable fossil shells. Mr. Brander took great pleasure in investigating these trea-

sures ; and his work upon them — now become very
rare, and wherein the greater proportion of the species
are figured — still remains the most valuable on the
subject. He died, it is said, in 1787, and bequeathed
his collections to the British Museum.

> *Fossilia Hantoniensia,* collecta et in Museo Bri-
> tannico deposita. London, 1766. 1 vol. thin 4to.
> with numerous well engraved figures.

BREHM, CH. LOUIS. — *Ornithology.*

A German clergyman, and an acute observer of birds.

1. *Materials for a History of Birds* (in German).
Neustadt, 1820—22. 2 vols. 8vo.

2. *Handbuch* der Naturgeschichte aller Vögel
Deutschlands. Ilmenau, 1831. 1 vol. 8vo. with 47
coloured plates.

BREYNIUS, J. P. — *Zoology.*

A physician of Dantzic. Born 1680, died 1764.

1. *Dissert. de Polythalamiis,* nova Testaceorum
Classe. Dantz. 1732. 4to.

2. *Historia Naturalis* Cocci Radicum Tinctorii.
Gelani, 1731. 4to.

BRISSON, MATHURIN JACQUES. — *Mammalogy and Ornithology.*

Mathurin Jacques Brisson, well known as an accurate describer of birds, was a native of Fontenay-le-
Comte, born on the 30th of April, 1723. He early
became acquainted with Réaumur, then prosecuting his
researches both in natural history and general physics;
and this acquaintanceship gave a direction to his future
pursuits. Réaumur employed him to take charge of
his museum, which was extensive; and also to assist

him in his various works. The influence of his patron,
in connection with his own great merits, raised him to
several stations of importance. He succeeded the abbé
Nollet in the chair of Physics in the College of Navarre.
He instructed the family of the French monarch in
physical science and natural history; and he was a
member both of the Academy of Sciences and of the Institute. He was the author of many treatises on elec-
tricity, aërostatics, specific weight of bodies, weights and
measures, and other similar subjects. It was, doubtless,
the facilities afforded by Réaumur's museum that led
him to undertake his various works on birds, which entitle him to hold an honourable place among naturalists.

Brisson died at Broissi, near Versailles, on the 23d
of June, 1806. An attack of apoplexy, some months
previous to his decease, completely destroyed his recollection of every thing, including even the French language; the only words he could recall to his memory
were a few provincial ones with which he had been familiar in his boyhood.* — J. D.

1. *Regnum Animale* in Classes novem distributum,
cum duarum primarum Classium, Quadrupedum sci-
licet et Cetaceorum, particulari Divisione, Latine et
Gallice. Paris, 1756. 4to. pp. 382. pl. 1.

2. *Le Règne Animal,* divisé en IX. Classes. Paris,
1756. 4to., containing only the Quadrupeds and Testacea.

3. *Ornithologie,* ou Méthode contenant la Division
des Oiseaux en Ordres, Sections, Genres, Espèces,
et leurs Variétés. Paris, 1769. 6 vols. 4to. In
this valuable work, the scientific ornithologist will
find recorded many of those genera which modern
systematists have only revived or modified. The
great value of the descriptions (which are both in
French and Latin), is their length and minute exactness; for, as the species were drawn and described
from the specimens in the cabinet of M. Réaumur,

* Further observations on this work will be found in our Preliminary
Discourse, page 78.

the author has judiciously refrained from quoting what others have said on their habits or economy. The plates, in general, are tolerably exact; but, from their not being coloured, cannot be depended upon in doubtful cases: the proportions of the figures are good, but the designs are made without taste or scientific knowledge.

4. *Ornithologia,* sive Synopsis Methodica sistens Avium Divisionem in Ordines, &c. Lugduni Batavorum, 1763. 2 vols. 8vo. This may be considered as a synopsis of the last work, with the descriptions abridged, and specific characters added after the manner of Linnæus.

BROCCHI, G.— *Conchology.*

This able Italian united the singularly opposite qualities of an excellent conchologist, and an eminent military engineer. He is stated by Cuvier to have died in Syria, in 1828, when in the service of the pacha of Egypt.

Conchiologis Fossilis Sub-Apennina. Milan, 1814. 2 vols. 4to.

BRODERIP, W. I.— *Conchology.*

An acute and most correct zoologist; whose descriptions are no less accurate than elegant, and who is peculiarly happy in describing the habits of animals. His various papers, chiefly relating to conchology, will be found in the Zoological Journal and Transactions.

BRONGNIART, ALEXANDRE. — *Zoology.*

Professor in the Faculty of Sciences in Paris. Born in 1770. An eminent naturalist.

1. *Essai d'une Classification* Naturelle des Reptiles. Paris, 1805. 4to.

2. *Histoire Naturelle des Crustacés Fossiles,* sur

les Rapports Zoologique et Géologique ; savoir, les
Trilobites, par A. Brongniart; et les Crustacés propre-
ment dits, par A. G. Desmarest. 1 vol. 4to. Paris,
1812.

BROUSSONET, P.-M.-A. — *Ichthyology.*

Pierre-Marie-Auguste Broussonet was the son of a
physician, and born at Montpellier on the 28th of February, 1761. He, likewise, was instructed in the healing art, and obtained his degree at the early age of
eighteen. He soon after removed to Paris, where he
employed himself principally in studying natural history.
He then visited England, and resided for some time in
the house of sir Joseph Banks. While in London, he
published a work on fishes, intitled *Ichthyologiæ Decas* 1*a*.
Lond. 1781. He also read a memoir on *Ophidium* to
the Royal Society, and became a member of that body.
On his return to Paris, he was patronised by Buffon,
who made him his assistant both in the College of France
and in the Veterinary School. In 1785, he was chosen
secretary to an Agricultural Society, established by Berthier de Sauvigni. About this time, he began to take a
part in the political proceedings which then convulsed
the country, and his life thereafter was of a very che-
quered description. On retiring to Montpellier, he was
arrested, but contrived to make his escape, and travelled
on foot, without money, and almost without clothes, over
the Pyrenees, botanising as he went, to Madrid, where
he was hospitably received by the botanists Ortega and
Cavanilles. He then embarked in an English vessel for
India, but was driven into Lisbon by stress of weather.
From this place, he found his way into Africa, where
he collected many plants, and transmitted specimens to
sir Joseph Banks, who had generously sent him a supply
of money in his distress. On the return of the emigrants
to France, he speedily revisited his native country;
and having succeeded in obtaining the appointment
of consul at Mogador, embarked for that place with

his family.   A relative, holding an influential station
in the government, at last obtained for him the situation
of Professor of Botany in the School of Montpellier; and
he ultimately settled in his native city.   He was cut off
by apoplexy on the 27th of July, 1807.   It is mentioned as a curious feature in his disease, that after the
first attack, he completely lost the recollection of proper
names and substances; while adjectives, whether French
or Latin, occurred to him as readily as before. — J. D.

   Ichthyologia, Fas. 1. royal 4to.   London and Paris,
1782.   The excellency of this work leads us to regret
that it was never continued.

Brown, Peter. — *Ornithology.*

A professional artist, of little merit as a describer.

   New Illustrations of Zoology.   London, 1776.
4to. coloured plates 50.   These plates are certainly
superior, on the whole, to those of Edwards, and
several are very good; the descriptions, however, are
so short as be nearly useless.

Brown, Dr. Patrick. — *General Zoology.*

An eminent physician, much attached to botany, who
for many years resided in Jamaica, and ultimately settled in Ireland, where he died.

   The Civil and Natural History of Jamaica, with
complete Linnæan Indices.   London, 1780.   1 vol.
folio, many plates.

Brown, Capt. — *Conchology.*

Late captain in the Forfar militia.

   1. *Elements of Conchology,* or Natural History of
Shells, according to the Linnæan System.   London,
1816. 8vo. p. 168. pl. 9.   The author observes, that
" the beauties of the Linnæan system will perpetuate

its pre-eminence." The figures are ill drawn and badly engraved.

BRUGUIÈRES, JEAN-GUILLAUME. — *Conchology.*

Jean-Guillaume Bruguières was born at Montpellier, in the year 1750. He first studied medicine, but ultimately abandoned it, and devoted himself entirely to natural history. He accompanied the expedition sent out by the French government in 1773, under the command of Kirguelin, to make discoveries in the South Sea, — acting as naturalist. On their return, the captain was tried for various misdemeanors, and imprisoned; and although he published a narrative of the voyage, Bruguières contributed nothing in his department. But the result of his researches appeared at intervals in separate memoirs inserted in the *Journal de Physique.* On returning to Montpellier, he happened to assist in an attempt to discover a bed of coal: the operations necessary for that purpose disclosed a great number of fossil shells; and the examination of these is said to have first inspired him with a desire to study the testaceous *Mollusca,* which he continued to do for a time with great diligence. He then went to reside at Paris, where he was employed in writing for the *Encyclopédie Méthodique.* He drew up the first volume of that elaborate and valuable work relating to the natural history of the *Vermes.* His method of classifying them is affirmed by Cuvier to be greatly superior, in many respects, to the arrangements of any of his predecessors in the same department; his descriptions are clear and minute; and many new species are introduced. When the well-known entomologist, Olivier, set out on his journey to the East, in 1762, Bruguières accompanied him. They visited Constantinople, the Grecian Archipelago, Egypt, Syria, Persia; residing for a considerable time at Teheran, and Bagdad; and returning by Asia Minor, Greece, and the Ionian Islands.*

* Olivier has published an account of his Eastern Journey (Paris, 1801—1804.).

Bruguières had been induced to undertake this journey, partly by the hope of improving his health ; but he continued so feeble all the time, that he could do little, notwithstànding his zeal, to make discoveries in natural history. He died, soon after landing at Ancona, on the 1st of October, 1799, in the 49th year of his age. He was an Associate of the Institute of France. A genus of plants (*Bruguiera*) has been named after him by M. du Petit-Thouars.

As a conchologist, he possessed very great merit. He separated several genera from those of Linnæus; while his accurate discrimination of species was no less remarkable than the clearness and precision of his descriptions. Unfortunately, he never completed more than the following volume: — *Dictionnaire des Vers.* Paris, 1792. 4to., forming a part of the voluminous *Encyclopédie Méthodique.* — J. D.

BRÜNNICH, MARTIN THOMAS. — *Ornithology.*

A Danish naturalist of great merit, and Professor in the University of Copenhagen.

1. *A History of the Eider Duck* (in Danish). Copenhagen, 1763. 1 vol. 8vo.

2. *Ornithologia Borealis.* Hafniæ, 1754. 8vo. p. 80. pl. 1.

3. *Entomologia* sistens Insectorum Tabulas Systematicas. Hafniæ, 1764. 1 vol. 8vo.

4. *Ichthyologia Massiliensis*, &c. Hafniæ et Leipsic, 1768. 1 vol. 8vo.

BUCHANAN, DR. FRANCIS HAMILTON. — *Ichthyology.*

A learned and accomplished Oriental traveller, and a most accurate zoologist and botanist. His writings evince great talent, and will ever remain a monument of critical skill and deep research. Besides his travels through the Mysore territories of India, he is chiefly

L

known for the most valuable account of the fishes of the East yet published. He died, we believe, in his native city of Edinburgh, in 1829.

Natural History of the Fishes of the Ganges. Edinburgh, 1822. 1 vol. 4to. and an atlas of plates. Although we cannot admit the principles of nomenclature, too frequently acted on in this volume, of coining pseudo-Latin names, out of barbarous Indian words, we esteem this the best work on ichthyology which has ever appeared in this country, and equal to those of any other. Under his former name of Buchanan, the author wrote a valuable paper on the new genus *Onchidium,* in the Linn. Trans. vol. v. p. 132. pl. 5.

BUFFON. — *General Zoology.*

George Louis le Clerc Buffon, who from his rank and talents exercised a considerable influence on the state of natural science in Europe, was a native of Montbard in Burgundy, born in September, 1707, the same year that gave birth to Linnæus, whose opposite he was in almost every respect. We possess but scanty records of his early life, but he is said to have shown great vigour of mind and aptitude for learning in his boyhood. The College of Dijon was the seminary where he was first placed, with a view of studying law; but for that he soon evinced an insuperable dislike, and devoted himself to more abstruse sciences, particularly mathematics and astronomy. In these, his attainments are allowed to have been considerable — nay, extraordinary; his attachment certainly was; he always carried a copy of his favourite Euclid with him as his pocket companion. At Dijon, he formed an acquaintanceship with a young nobleman of the name of Kingston, then on his travels through Europe, and it was arranged that Buffon should accompany him and his tutor on their way southwards. They visited Italy, and various other parts of Southern Europe. During his absence, his mother (whose maiden

name was Merlin) died ; an event by which he came
into possession, even before attaining his majority, of
an income of nearly 12,000*l.* a year. He afterwards
visited Paris and England ; and about the age of twenty-
five returned to his paternal residence, and settled there ;
distributing his time in such a manner as to embrace a
regular and systematic course of study in the sciences
generally (more particularly natural history), and po-
lite literature. These studies were chiefly carried on
in a pavilion in his garden, which prince Henry of
Russia called the cradle of natural history ; and Rous-
seau, before he entered it, used to fall down on his
knees and kiss the threshold. The first work Buffon
published, was a translation of Hales's " Vegetable Sta-
tistics ;" and soon after that an edition of Newton's
" Fluxions." His great ambition, at the commencement
of his authorship, was to acquire a pure and elegant
style, copious, popular, and attractive,—an object in
which he is well known to have succeeded in no ordinary degree. The interest he felt in the subject of the
first-mentioned work, led him to make various experiments to determine the relative strengths of different
woods used for public purposes : about the same time,
also, an extensive series of experiments was undertaken to
show what effects might be produced by burning-mirrors
of particular form and arrangement ; a subject suggested
to him by what is recorded in history, of Archimedes
setting the Roman fleet on fire by such means. The result showed the perfect possibility of such an occurrence.

In 1739, Buffon succeeded M. du Fay as Intendant to
the Royal Garden and Cabinet, an office which induced
him to devote his attention chiefly to natural history.
He entered upon the pursuit, to which he had always
been strongly inclined, with his characteristic ardour,
and soon formed the plan of a great work, which was
to embrace not only the history of the earth, but every
department of the animal kingdom. It was long, however, before the first zoological volume appeared, although he is said to have employed fourteen hours daily

in study ; the whole of the first edition of the work
was not completed till 1767. It amounted to 15 quarto,
and 31 octavo volumes. M. Daubenton assisted him,
particularly in the anatomical department. This portion
of the work included the quadrupeds only ; that com-
prising the birds did not appear till 1771, and in preparing it he availed himself of the aid of M. de Montbeillard and the abbé Beron. A natural history of
minerals, and several supplementary volumes, were afterwards added, so that the entire work ultimately amounted
to 38 quarto, and 62 octavo volumes. Numerous editions are before the public, and it has been translated
into most of the languages of Europe.

In 1752, Buffon was married to mademoiselle St.
Berin, an event by which his domestic comfort was
very much increased. His fame was now widely spread,
and honours from various quarters conferred on him.
In 1771, his estate was raised to the rank of a compté
by Louis XIV. ; and the monarch still further evinced
his favourable regard, by inviting Buffon to Fontainbleau,
and offering him the office of administrator of the forests
throughout his dominions. This appointment, however, the Count thought proper to decline. Notwithstanding the irregularities in which he had indulged
in his younger days, and, indeed, at intervals during
the greater part of his life, his health continued un-
impaired till his 72d year. About that time he became afflicted with the stone, a disorder from which he
suffered great agony for the remainder of his days.
No consideration could prevail on him to submit to an
operation, although he was assured that, in his case,
that could be performed with comparative safety, and
would afford immediate relief. During all his sufferings,
his studies were carried on with little interruption ; the
habit having become so confirmed by long usage, that
it was easier to continue than to lay them aside. In
this state he remained for nearly eight years, during
which his disorder reached such a height as to occasion
extreme torture. His death took place on the 16th of

April, 1788; he was, therefore, in his 81st year.—
J. D.

1. *Histoire Naturelle des Oiseaux.* Paris, 1770—
83. 9 vols. 4to., containing 973 plates of Birds, and
35 of Insects. With this was published the celebrated Planches Enluminées.

2. *Histoire Naturelle, générale et particulière,* avec
la Description du Cabinet du Roi. Paris, 1749—89.
36 vols. 4to.; of which 3 are on general subjects,
12 on quadrupeds, 7 supplementary to the former
subjects, 9 on birds, and 5 on minerals. The succeeding editions of this great work are numerous. The
principal are —

Histoire Naturelle, générale, et particulière. Paris,
1769—79. 52 vols. 12mo., with the plates reduced.

Histoire Naturelle, &c. Deux Ponts, 1785. 43 vols.
12mo., of which the plates were coloured.

3. *Histoire Naturelle,* &c., augmentée de Notes, et
rédigée par C. S. Sonnini, &c. 1807. 127 vols. 8vo.
The plates of all these editions, with very few exceptions, are all copied one from the other. The designs,
particularly in the ornithological department, are stiff,
unnatural, and frequently glaringly incorrect; in
fact, many can hardly be recognised. It is really surprising, with so many really good zoological painters in
France, that such contemptible prints should still be
affixed to the numerous editions of a work which the
French boast so much of.

BURROWS, THE REV. E. J.—*Conchology.*

A minister of the established church.

Elements of Conchology, according to the Linnæan
System; illustrated by 28 Plates, engraved by Heath,
from Drawings by the Author. London, 1818. 1 vol.
8vo. pp. 245. pl. 28. The beauty of the plates will
commemorate this work, more than the system which
the author has advocated. The figures are mere outlines, yet drawn with such perfect truth, and delicacy

of touch, that, for this particular style, they may be consulted as models.

CARENA, GIACINTO. — *Malacology.*

Late Professor of Natural History at Turin : known only by his *Monograph of the Genus Hirudo,* in the 25th volume of the " Transactions of the Academy of Turin." 4to. 1820.

CARMICHAEL, CAPT. — *General Zoology.*

A British officer, whose *Essay on the Natural Productions, &c. of Tristan d'Acunha* (Linn. Trans. vol. xii.) is not merely the only account we possess of that island, but is a model for treating such a subject. It contains the description and figures of several new and interesting fishes.

CATESBY, MARK. — *General Zoology.*

The first author who commenced the folio system of coloured plates in natural history on an expensive plan, now so much pursued in purely illustrative works.

Natural History of Carolina, Florida, and the Bahama Islands. Large folio, 2 vols. with 220 coloured plates. London, 1750. The text is in French and English, and the figures very good for the period of their publication.

CAVOLINI, P. — *Malacology.*

A Neapolitan physician, attached to natural history.

1. *Memorie* per servire alla Storia de' Polipi Marine. Napoli, 1785. 4to.
2. *Sulla Generazione dei Pesci* e dei Granchi. Napoli, 1787. 4to.

CETTI, FRANCESCO. — *Ornithology.*

Storia Naturale di Sardegni. Sassari, 1774—77.
4 vols. 12mo. Of this rare work we have only seen
one volume.

CHABERT. — *General Zoology.*

Director of the Veterinary School at Alfort.
Traité des Maladies vermineuses dans les Animaux.
Paris, 1782. 8vo.

CHABRIER, J. — *Entomology.*

A philosophic entomologist, now alive, and residing
in Paris.

Essai sur le Vol des Insectes, et Observations sur
quelque Parties de la Mécaniques des Mouvemens
progressifs de l'Homme, et des Animaux Vertébrés.
Paris, 1822. 4to. pl. 14.

CHARPENTIER, TOUSSAINT DE. — *Entomology.*

Horæ Entomologiæ adjectis Tabulis novem coloratis
Wratislaviæ.   Paris, 1825.   1 vol. 4to. coloured
plates 9.

CHEMNITZ, J. J. — *Conchology.*

An assiduous conchologist, whose name is usually associated with that of Martini, because both cultivated the
same branches of natural history, and their most important
works were joint productions. Jean Jerome Chemnitz
was born at Magdeburg, on the 10th of October, 1730;
and died on the 12th of October, 1800, at Copenhagen.
He was an ecclesiastic, and most of his works relate to
theological subjects : but he was much devoted to the
study of nature, and greatly excelled in his knowledge

of conchology, the department to which he chiefly applied himself. On this subject he has published several Memoirs in the " Beschdefligungem Naturforschinder Freunde zu Berlin," and in another work, entitled " Naturforscher." Both his professional and scientific knowledge were happily combined in a work published at Nuremberg in 1760, under the title of *Kleine Beytraage zur Testaceotheologie, oder zur Erkenntniss Gottes aus der Conchylien.* But his principal claim to the gratitude of naturalists is founded on his being the continuator of the great work on conchology begun by Martini. The latter had completed three volumes before his death, when it was taken up by Chemnitz, who published eight others between the years 1779 and 1796. This still continues to be a standard work of reference, although the arrangement is objectionable, and the figures are by no means good.—J. D. (See MARTINI.)

CLAIRVILLE.—*Entomology.*

A learned entomologist, of English extraction, long resident in Switzerland.

Entomologie Helvétique. Zurich. Vol. I. 1798.; Vol. II. 1806. The descriptions and plates have equal merit.

CLERK, CHARLES.—*Entomology.*

An eminent zoological painter, patronised by Linnæus.

1. *Aranei Suecici* Descriptionibus et Figuris illustrati. Holmiæ, 1757. 4to. pp. 154. col. pl. 6. This work, valuable for its figures, is rare.

2. *Icones Insectorum* rariorum. Holmiæ, 1759 —1764. The plates are highly finished, and are particularly quoted by Linnæus and Fabricius. The book, however, is even of greater rarity than the last.

CLUSIUS, C. — *Zoology and Botany.*

One of the early fathers of botany. Born in 1526, and died in 1609. He was physician to the emperor, and subsequently Professor of Botany at Leyden.

Exoticorum Libri X. Anvers, 1605. 1 vol. folio.

COLONNA, FABIUS. — *Zoology.*

A learned physician of the 16th century, whose writings are still valuable. He was born in 1507, and died in 1660.

1. *Aquatilium* et Terrestrium aliquot Animalium, aliarumque Naturalium Rerum Observationes. 1616. 4to.

2. *De Purpurâ,* &c. Romæ, 1616. 4to.

COMMERSON, P. — *Voyager and Naturalist.*

Philibert Commerson was a celebrated naturalist and traveller, distinguished for his intrepidity and ardent thirst for knowledge. He was a native of Chatillon-les-Dombes, born on the 18th of November, 1727. Designed for the medical profession, he repaired, at the age of twenty, to Montpellier, then celebrated as a school of medicine, and which has produced, as may be seen from these biographical notices, no small number of eminent naturalists. In the latter capacity, he soon attracted the regard of Linnæus, who invited him, in the name of the queen of Sweden, to collect and describe the fishes of the Mediterranean. This congenial task he entered upon with the greatest ardour, and completed it to the entire satisfaction of his patrons, — making a valuable contribution to this difficult branch of natural history. He obtained his degree of M.D. in the year 1755. After travelling some time in Switzerland and Savoy, to gain an acquaintance with the plants indigenous to these countries, he went to reside at Chatillon, where he remained for eight years. While there, he employed a portion of his time in forming a botanical garden. By the representations of his friend Lalande, he was induced to visit Paris, where there were more frequent opportunities of rendering his scientific knowledge of service both to himself and the public. When the expedition under Bougainville was determined on, Commerson was appointed to accompany it as naturalist. He accordingly set sail in the beginning of the year 1767; and after visiting many parts of the globe, went to the Isle of France, where he was directed to remain for a time, in order to investigate the natural productions both of that place and of Madagascar. It was in the former of these islands that his death took place, in 1773. Only fragments of his writings have been published; but his papers, drawings, and collections are numerous, and, being carefully preserved in the Jardin du Roi, they are yet made available to naturalists.—J. D.

Cooke.— *Ornithology.*

Description of the Whistling Swan, and of the peculiar Structure of its Trachea. London, 1823. folio, pp. 11. pl. 2.

Coquebert, A. J.—*Entomology.*

An assiduous entomologist of Rheims.

Illustratio Iconographica Insectorum, quæ in Musæis Parisinis observavit J. Ch. Fabricius. 4 Nos. in 4to. Paris, 1799—1804. The plates are well filled, but poorly executed; yet, as most of the figures are magnified, the subjects are easily recognised, and thus afford valuable references.

Costa, Emanuel Mendez da.— *Conchology.*

A Portuguese merchant, long resident in London, and an ardent lover of conchology.

1. *Historia Naturalis Testaceorum Britannicæ;* or, British Conchology; containing the Natural History of the Shells of Great Britain and Ireland, in English and French. London, 1778. 4to. pp. 254. col. pl. 17. The figures are by Mazell, and are very good.

2. *Elements of Conchology,* or an Introduction to the Knowledge of Shells. London, 1776. 8vo. pp. 318. pl. 7.: in some few copies, the plates are coloured. He also commenced, with Humphrey, the publication of a General Conchology, in folio numbers, which was to contain figures of every known shell. The work was given to the public anonymously; and the figures, engraved by Mazell, were certainly the best that had yet appeared: but few subscribers were found to embark in such a vast undertaking; so that after five or six numbers had appeared, its further publication was relinquished.

COUCH, JONATHAN. — *Conchology.*

An able and most assiduous ichthyologist, now living on the coast of Cornwall. He has written a paper on the rare fishes of that coast, in the Linn. Trans. vol. xiv. p. 69., and some others in the " Magazine of Natural History," besides contributing largely to the " British Ichthyology" of Mr. Yarrell.

CRAMER, PIERRE. — *Entomology.*

A wealthy merchant of Amsterdam, and member of its scientific institutions.

Papillons Exotiques des Trois Parties du Monde, l'Asie, l'Afrique, et l'Amérique. Amst. 1779— 1782. 4 vols. 4to. plates coloured, 400. This is the most complete and the most valuable work ever published on exotic lepidopterous insects. The text is destitute of solid information or of useful re- marks; but the figures, though faulty and inaccurate in their outlines, are well coloured, and can immediately be

recognised. A supplementary volume was published
by Stoll, of much more scientific value, as it contains figures of the larvæ and pupæ of many rare
species found in Surinam and Brazil.

CREUTZER, CHRISTIAN. — *Entomology.*

An entomologist of Germany, whose papers appear to
have been published under the title of

   Entomologishe Versuche, or Entomological Essays.
Vienna, 1799. 1 vol. 8vo. with coloured plates.

CUBA, J. — *General Zoology.*

We introduce this author into our list, not from any
scientific value attached to his work, but because it is the
earliest and most curious book on natural history published after the dark ages. There is an immense number
of singular, and often ludicrous, figures on wood; giving
us the earliest specimen, perhaps, of this art being applied
to the delineation of animals.

   𝔒rtus 𝔖anitatis. 2 vols. small folio, with a Treatise *De Urinis.* Moguntiæ, J. Meydenbach, 1491.
There is another edition, printed at Antwerp in 1517,
in one folio-volume. (*Bohn's Cat.*)

CURTIS, W. — *Entomology and Botany.*

An assiduous botanist, and one of the principal London nurserymen and florists of the period.

   1. *Fundamenta Entomologiæ,* or an Introduction
to the Knowledge of Insects. Lond. 1772. 8vo.
pl. 2.

   2. *A short History* of the Brown-tailed Moth.
London, 1782. small 4to. The caterpillar of this
moth appeared in such amazing myriads in the environs of London in this year, that the public appre-
hended the approach of a plague or famine. To
quiet these fears, Mr. Curtis very judiciously pub-

lished this little tract, explaining the nature of the
insect, and the most effectual mode of destroying it.

CURTIS, JOHN. — *Entomology.*

A celebrated engraver and draftsman of natural history, and an acute naturalist; justly considered the first
entomological artist in existence; whose admirable and
exquisite style of representing insects was soon taken
up by the Continental artists.   He is the possessor of
a most extensive cabinet.

   British Entomology, or Illustrations and Descriptions of the Genera of Insects found in Great Britain
and Ireland; containing coloured Figures from Nature of the most rare and beautiful Species, and of
the Plants upon which they are found.   Royal 8vo.
The first number published in 1824, the last terminated in 1840.   Mr. Curtis is also the author of one
or two other small works upon British insects, which
we have not seen; and of an interesting paper on the
Elater Noctilucus, in the last volume of the " Zoological Journal."

CUVIER, BARON. — *General Zoology.*

The name of Baron Cuvier may be regarded as marking an era in the history of natural science, in consequence of the flood of light he threw upon the subject,
by his unequalled knowledge of comparative anatomy,
and the study of the internal organisation of animals in
connection with their external forms.   He was born at
Montbéliard, on the 23d of August, 1769, and baptised
by the name of George Leopold Chrêtien Frederick
Dagobert Cuvier.   His father was an officer in a Swiss
regiment in the service of France, and was ultimately
appointed commandant of the artillery at Montbéliard,
with a small pension.   After some rudimentary instructions from a most tender and affectionate mother,
young Cuvier, at the age of ten, was placed in the public

gymnasium, where he greatly distinguished himself, and
where his taste for natural history first showed itself
by his making the works of Gesner and Buffon his
favourite study. It was afterwards arranged that he
should be sent to a school at Tubingen, in order to
prepare him for the church, as that seemed to his
parents the only channel through which they could
procure him preferment ; but prince Charles of Wur-
temberg happening to visit Montbéliard, and becoming
acquainted with his abilities and circumstances, took the
youth (than fourteen years of age) along with him to
Stutgard, and enrolled him in the university of that
place at his own expense. Here his studies had a very
wide range, and he highly distinguished himself in every
department he entered upon, carrying off the highest
prizes, and having an order of chivalry conferred upon
him. On leaving the university, it was designed that
he should enter some branch of the administration ; but
an appointment of that nature being not readily obtained,
he became a tutor, at the age of nineteen, in the family
of Count d'Hericy, residing near Caen, in Normandy.
Here he resided nearly seven years, devoting much of
his leisure time to the anatomy of the *Mollusca*, and
others of the lower animals ; here also he became acquainted with M. Tessier, by whom he was introduced
to many of the leading naturalists of the day. By the
advice of his friends, he removed to Paris in 1795.
Soon after his arrival, he was appointed a professor in
the Central School of the Pantheon ; and, in the same
year, assistant to M. Mertrud, who occupied the chair
of comparative anatomy in the Jardin des Plantes.

Thus placed in a situation for which his acquirements and the natural bent of his mind admirably
qualified him, a brilliant prospect opened before him ;
but into the details of his long and active career we
cannot here enter. All that our limits permit us to
attempt, is to notice the principal events of his life, and
the various official situations which he held.

When the National Institute was established, he became one of its earliest members; he was solicited to accompany the savans who attended the expedition to Egypt, which, however, he declined; and when Napoleon assumed the title of President of the Institute, Cuvier acted as secretary. His next appointment was to the perpetual secretaryship of the Institute in the class of Natural Sciences, with a salary of 6000 francs; and in that capacity we are indebted to him for many admirable reports on the progress of natural knowledge, and historical sketches (*éloges*) of deceased members. In 1800, he was nominated successor to M. Daubenton, the accomplished coadjutor of Buffon; soon after this appointment, he married the widow of M. Duvaucel. An important duty devolved on him, when he was appointed one of the counsellors of the Imperial University; as he was obliged to visit both Italy and Holland, for the purpose of superintending the establishment of universities and academies. Enjoying the favour of the emperor, at this time the dictator to Europe, honours were accumulated on the head of Cuvier: the lucrative office of Maître des Requêtes was assigned to him; and in 1814 he was raised to the rank of counsellor of state. The latter office he continued to hold under Louis XVIII., during that monarch's temporary occupation of the throne of France; and even on Napoleon's return from Elba, his connection with the universities still continued. After the second restoration of the Bourbons, Cuvier's influence was as great as at any former period; it was augmented under Charles X.; and after the last revolution, he was created a peer of France, and still further dignities were intended for him.

Notwithstanding Cuvier's multifarious official duties in his own country, and his habits of diligent study, he found time to pay a visit to England, where he met with a most friendly reception. On his return to France, he was created a baron by Louis XVIII.; and he may be considered, about that period, as having reached the height of his power and fame. His saloons were eagerly resorted to by the most distinguished literati of Europe,

naturalists and travellers from every clime, as well as by
politicians and diplomatists. The only interruption to
his happiness was the death of his four children, the
eldest of whom had reached the age of womanhood.
Just before the outbreak of the revolution produced by
the ordinances of Charles X., Cuvier had set out to pay
a second visit to London : intelligence of that singular
event, which reached him before he arrived at Calais,
did not make him forego his purpose ; and on his return to Paris, he found himself, notwithstanding the
change of dynasty, in full possession of all his offices,
honours, and dignities. The first public indication that
his active and useful life was drawing to a close, appeared at the termination of his course of lectures on
the history of the natural sciences, in the College of
France, which, from the tone in which it was conceived
and delivered, was regarded as a kind of farewell to his
pupils, and seemed " prophetic of his end." He was, in
fact, seized with paralysis shortly after, and expired in
a few days.

For an enumeration of baron Cuvier's works, and the
dates of their publication, we must refer to the list appended to this sketch of his life. Besides his larger
works, a very great number of insulated memoirs on various subjects emanated from his pen; so early as 1818,
these amounted to no fewer than 127, many of them
of great length.   The first he appears to have published,
was in 1792, while he resided at Caen ; the subject was
L'Anatomie de la Patelle. The famous *Recherches sur
les Ossemens Fossiles des Quadrupèdes,* were at first published in detached parts in the " Annales du Muséum,"
and subsequently collected into four quarto volumes.
In the *Règne Animal distribué d'après son Organisation,*
he was assisted by many of his colleagues; and the entire department relating to insects is from the pen of
Latreille.   M. Valenciennes was his collaborateur in his
work on ichthyology.   By these valuable aids, he was
able to perform the numerous political and administra-
tive duties belonging to his offices, and to give a uniform

character with his own to the labours of his assistants. The valuable library he accumulated, amounting to 19,000 volumes, was purchased by the government, and divided between various schools of medicine, law, and natural history.—J. D.

1. *Tableau Elémentaire* de l'Histoire Naturelle des Animaux. Paris, 1798. 8vo.

2. *Leçons d'Anatomie Comparée*, récueillées et publiées par MM. Duméril et Duvernoy. Paris, 1800—1805. 5 vols. 8vo.

3. *Recherches sur les Ossemens Fossiles* des Quadrupèdes. Paris, 1812. 4 vols. 4to. An enlarged edition of this most valuable and erudite work has since been published.

4. *Ménagerie du Muséum* d'Histoire Naturelle, par MM. Lacepède, Cuvier, et Geoffroy ; avec des Figures peintes par Maréchal. Paris, 1804. 2 vols. 8vo. There is another edition in large folio.

5. *Mémoires* pour servir à l'Histoire et à l'Anatomie des Mollusques. Paris, 1816. 4to. plates.

6. *Le Règne Animal* distribué d'après son Organisation. Paris, 1817. 4 vols. 8vo. An English translation, considerably augmented, has been terminated by Mr. Griffith "and others," accompanied by plates, many of which boast the name of T. Landseer.

CUVIER, FREDERIC.—*Mammalogy.*

Late inspector of the French Academy, and curator of the menagerie of the Museum of Paris. Born at Montbéliard, in 1773 ; died in 1838. In conjunction with Geoffroy St. Hilaire, he has published—

1. *Des Dents des Mammifères,* considérées comme Caractères Zoologiques. Paris, 1825. 1 vol. 8vo. with 100 plates of the teeth of different genera. Little or no notice is taken of any other of the characteristics belonging to these genera ; which, added to the omission of their classic names, renders this work

M

of far less utility than it otherwise would have been. It belongs more, indeed, to comparative anatomy, than to zoology.

2. *Essai sur la Domesticité des Mammifères*, précédé de Considérations sur les divers Etats des Animaux, dans lesquels il nous est possible d'étudier leurs Actions. Paris, 1826. 4to.

3. *De la Sociabilité* des Animaux. Paris, 1824. 4to.

4. *Observations* sur la Structure et le Développement des Plumes. Paris, 1826. 4to. We believe these three last pamphlets were originally inserted in some of the French scientific Transactions, and a few copies afterwards printed off and sold separately.

CYRILLUS (OR CIRILLO), DOMINICO. — *Entomology.*

A learned physician of Naples, who fell a victim to the fury of the revolution which disgraced that city in 1796.

Entomologiæ Neapolitanæ Specimen. Napoli, 1787. 1 vol. folio, col. pl. A work of considerable merit, and of great rarity.

DAHL, GEORGE. — *Entomology.*

An esteemed German entomologist.

Coleoptera and Lepidoptera, Catalogue of. Vienna, 1823. 1 vol. 8vo. We can find no other account of this work than the foregoing, given by Cuvier and Percheron.

DAHLBOM. — *Entomology.*

1. *Clavis* Novi Hymenopterorum Systematis. Holmiæ, 1825. With a coloured plate.

2. *Prodromus* Hymenopterologiæ Scandinaviæ. Lundæ, 1836.

DAHLMAN, JOHN W.—*Entomology.*

A learned and acute entomologist; late director of
the Royal Museum at Stockholm, his chief works are,

1. *Analecta Entomologica.* Holmiæ, 1824. 1 vol.
4to. plates.

2. *Prodromus* Monographiæ Castniæ, Generis
Lepidopterorum. Holmiæ, 1825. 4to. 1 plate.

3. *Monograph* of the Chalcidites, or Insects of the
Family of *Pteromalini.* Stockholm, 1820. 8vo.

4. *Synopsis* of the Lepidopterous Insects of Sweden, in the Stockholm Transactions. 1816.

5. *Ephemerides Entomologicæ.* Holmiæ, 1824.
1 vol. 8vo.

6. *Memoir* on certain Ichneumonides. Stockholm,
1826. 1 vol. 8vo.

7. *Prodromus* Monog. Generis Lepidopterum.
Stock. 1828. 4to. with coloured plates.

8. *Insectorum* novo Genera. Holmiæ, 1819. 8vo.

DALYELL, J. GRAHAM.—*Malacology.*

Observations on various interesting Phenomena of
the Planaria. Edinburgh, 1814. 8vo.

DANIELLS, SAMUEL.—*Animal Painter.*

An admirable landscape and animal painter, who
visited India, and travelled in Southern Africa with
Mr. Barrow. Besides the scenery, &c. of those countries, he has figured many of the rare and interesting
animals in a picturesque style.

1. *Ceylon.* Picturesque Illustrations of the Scenery, Animals, and native Inhabitants. London,
1808. folio.

2. *Southern Africa.* Sketches, representing the
native Tribes, Animals, and Scenery. London,
1820. Royal 4to. This latter is one of the most

M 2

beautiful books, in this branch, existing ; the sketches
are very masterly.

DAUBENTON, LOUIS J. M. — *General Zoology.*

This skilful coadjutor of Buffon, was a native of the
same place (Montbar, in Burgundy) as his more distinguished cotemporary, and born on the 29th of May, 1716.
He was acquainted with Buffon from his childhood;
and the more advanced age of the latter, and superior
worldly advantages, placed him almost in the relation of
a patron to Daubenton from the first. Daubenton was
sent to Paris to study theology, which he did with much
reluctance, gratifying himself for this restraint upon his
inclinations by studying medicine and anatomy in secret. On the death of his father, the latter branches
obtained the whole of his attention, and he took his degree at Rheims in 1741. He then returned to his
native place with a view to practise, and here his intimacy with Buffon was again renewed. Buffon took
him to Paris in 1742, and a few years afterwards gave
him the appointment of curator and demonstrator to the
Cabinet of Natural history. In this situation he spent
about fifty years of his life in great comfort and tranquillity. He laboured at his favourite pursuits with
indefatigable zeal ; and the number of interesting and
valuable facts in natural history, which he has made
known, is surprisingly great. Much of the most valuable matter in Buffon's great work was contributed by
Daubenton. All that relates to anatomy in the first
five volumes of the quarto edition is the result of his
labour. Cuvier regards this as essentially necessary to
enable the reader to understand the text; and yet it
was omitted in an edition of the work subsequently
published by Buffon. Daubenton's contributions did
not extend beyond the volumes mentioned ; in the history of birds, Buffon availed himself of the assistance
of Montbeillard and Beron. Notwithstanding the duties of his situation, which were numerous and laborious,

Daubenton was the author of many works, and a considerable number of memoirs in the "Transactions of the Academy of Sciences," in the "Encyclopédie Méthodique," &c.

As a public teacher, he was fully as successful as in his capacity of author. Cuvier informs us that he was the first who gave lectures on natural history in France by public authority. The subject had been formerly introduced in a subordinate sense, as an accessory branch of medical study. But, at his solicitation, one of the chairs in the College of France was converted into a chair of natural history, and he was appointed to fill it. This was in the year 1778. His other official stations were, the professorship of mineralogy in the Museum of Natural History, which he continued to hold till his death; and that of rural economy in the Veterinary School of Alfort. It was probably his connection with that institution which led him to make such exertions to introduce and propagate the Spanish breed of sheep in various parts of France, from which he anticipated much benefit in an economical point of view. He wrote several treatises on the subject, and, among others, one of a practical kind, intitled " Instructions to Shepherds."

Devoted to the tranquil pursuits of science, and naturally of a cheerful temperament, he enjoyed a large share of happiness during his long life. His wife, by whom he had no children, was of a literary turn, and published a romance under the name of *Zélie dans le Désert*, which is favourably spoken of.

A considerable inroad was made on his ordinary habits, and simple and uniform mode of life, when he was nominated a member of the senate in 1799. Indeed, the change was so great as materially to affect his health; and he had not attended many sittings of that assembly, when he was seized, during one of them, with an attack of apoplexy, which proved fatal. This happened on the 31st of December, 1799; he was consequently in his 83d year. — J. D.

DAUDIN, FRANÇOIS MARIE. — *Ornithology.*

François Marie Daudin was born at Paris, towards
the close of the 18th century. In his youth, he
was affected with some disorder which deprived him of
the use of his legs ; and he relieved the confinement to
which this calamity subjected him, by studying physics
and natural history. The latter soon became his chief
occupation ; and, during his short life, he produced a
considerable number of works,— none of which, however,
enjoy a very high degree of reputation. It was partly,
perhaps, the consequence of his bodily infirmity, that
he had but little personal acquaintance with the objects
he describes, most of his information being derived
from books. Several of his memoirs first appeared
in the *Magazin Encyclopédique* and the *Annales du
Muséum d'Hist. Nat.*, and they were afterwards collected and published in an octavo volume (Paris, 1800.),
under the title of " Récueil des Mémoires et des Notes
sur des Espèces inédites ou peu connues de Mollusques
et de Zoophytes." This was followed by a work on
quadrupeds and birds ; and a " Traité Elémentaire et
complèt d'Ornithologie." The best and most important
of his productions is a " Hist. Nat. des Reptiles," which
was written for Sonnini's edition of Buffon, and
amounts to eight octavo volumes. He shows, in this
work, a considerable acquaintance with the anatomy of
this class ; the genera are founded on stable characters ;
many new species are described ; and the history of
those previously known, is accurately detailed. A little
work on frogs and the allied tribes was among the last
of his productions. All of these are illustrated by numerous coloured plates, engraved from drawings made by
his wife. She is described as having been a very amiable
person : she was carried off by consumption, and her
husband survived her only a few days.— J. D.

1. *Traité Elémentaire* et complèt d'Ornithologie.
Paris, 1800. Two volumes (in quarto) are all that

were published of this work. Cuvier terms it a poor
compilation; but yet it is, in many respects, valuable,
as being the first book in which the discoveries of
Le Vaillant were put into a systematic form; or, in
other words, received specific names in Latin.

2. *Histoire Naturelle des Reptiles.*   Paris, 1802
—3. 8 vols. 8vo.   This work forms a part of Sonnini's edition of Buffon; but was also published
separately.

3. *Hist. Nat. des Rainettes,* des Grenouilles, et des
Crapauds. Paris, 1803. 1 vol. 8vo. coloured plates.

DEJEAN, COUNT.— *Entomology.*

A distinguished general in the wars of France; yet
still more celebrated for his devotion to entomology,
when no longer called upon to serve his country in the
field.   His collection is one of the most celebrated in
Europe; and he has studied it with profound attention,
and lasting benefit to his favourite science.

1. *Species Général* des Coléoptères. Paris, 1825—
31. five volumes of which, in thick octavo, have now
been published.

2. *Catalogue* de la Collection des Coléoptères de
M. le Conte Dejean. Paris, 1821. 1 vol. 8vo.

3. *Histoire Naturelle* et Iconographie des Coléoptères d'Europe; by M. Latrielle and Count Dejean.
Paris, 1822. Published in numbers, but not yet completed.

DELAUZE, M.

Although not a zoologist, this writer has a claim to
be admitted in the present list, from having made us
acquainted with the establishment of the French Museum. His work, originally written in French, has been
translated into English by A. Royer.

History and Description of the Museum of Natural History, and Royal Garden of Plants, at Paris,

with a full Account of the several Collections therein contained; composed .by Order of the French Government, from the Notes furnished by the Professors and Administrators of the Museum. London. 2 vols. 8vo. with plates.

DESHAYES, G. P.—*Conchology.*

An able and zealous conchologist, intimately acquainted with the fossil shells of France, who is now editing the new edition of Lamarck's great work.

Description des Coquilles Fossiles des Environs de Paris. 4to. Paris, 1824—1833. Published in numbers, with 4 plates in each. We know not if it terminated at, or extended beyond, the thirty-first.

DESMAREST, ANSELME-GAÉTAN.—*Ornithology.*

Professor of Zoology in the Veterinary School of Alfort. Besides being the author of several valuable contributions to the *Nouveau Dictionnaire d'Hist. Nat., Bulletin des Sciences,* &c., he has published a separate work on the

1. *Histoire Naturelle des Tangaras,* des Manakins, et des Todiers. Paris, 1805. folio. The figures are by mademoiselle Pauline de Courcelles, pupil of the celebrated Barraband : they are all beautiful, and many faultless ; but in general there is a want of life, of balance, and of variation in the attitudes. The text is valuable, and replete with information.

2. *Mammalogie,* ou Description des Espèces de Mammifères. Paris, 1822. 1 vol. 4to.

DICQUEMARE, THE ABBÉ JACQUES F.—*Malacology.*

We can find no other account of this naturalist, or of his works, than the following notice by Cuvier. " Dicquemare, a naturalist of Havre, was born in 1734, and died in 1789. An indefatigable observer, and author of various memoirs on the *Zoophyta* and *Mollusca* in the (French) *Philosophical Transactions, Journal de Physique*, &c."

DONATI, VITALIAN. — *Zoology.*

A physician of Padua, born in 1713. He was subsequently employed by the king of Sardinia, to travel for scientific information; and was shipwrecked on his return from Egypt, in 1762.

Natural History of the Adriatic, published in Italian. 1 vol. 4to. Venice, 1750.

DONOVAN, EDWARD. — *General Zoology.*

A laborious writer on natural history. Great labour has been bestowed upon the colouring of the plates he published, which renders his works expensive. The figures, for the most part, are destitute of grace or correctness, excepting, indeed, such as relate to entomology, most of which are faithful. The text is verbose, and not above mediocrity.

1. *Quadrupeds*, Natural History of British. 3 vols.

2. *Birds*, Natural History of British. London, 1794—1818. 10 vols. royal 8vo.

3. *Fishes*, Natural History of British. London, 1802—1808. 5 vols. 8vo.

4. *Shells*, Natural History of British. London, 1779. 5 vols. 8vo.

5. *Insects*, Natural History of British. London, 1792—1809. 15 vols. 8vo.

6. *Illustrations of Entomology*, including the Insects of China, India, and New Holland. London, 1799 —1805. 3 vols. 4to.

7. *Instructions* for collecting and preserving Subjects of Natural History, &c. London, 1794. 8vo.

8. *The Naturalist's Repository*, or Miscellany of Exotic Natural History. 5 vols. 8vo. London, 1834.

DRAPARNAUD, JACQUES PH. RAIMOND. — *Conchology.*

This zealous and able naturalist, whose early death must be lamented as a great loss to science, was born at Montpellier, on the 3d of June, 1772. His full name was Jacques Philippe Raymond Draparnaud. He was more remarkable than any of his youthful companions, in his love for study, and in an extraordinary aptitude for acquiring languages. His parents intended him for the bar; but he was anxious to indulge his taste for natural history, and medicine was chosen as the profession most compatible with such pursuits. After attending the College of Soreze for some time, he was appointed to the chair of general grammar in the Central School of the department of Herault. When the chair of natural history in the same establishment became vacant, he was transferred to it. A more important charge, however, was offered to him in 1802, and he became Professor of Natural History in the School of Medicine at Montpellier, with the title of Conservator of the Museum. This office he held but for a short time; for the new regulations promulgated respecting medical schools, restricted the field of his exertions so much that he resigned. Indeed, this step had become almost necessary on other grounds; for the seeds of consumption were inherent in his constitution, and his strength was rapidly giving way. He did, in fact, survive his resignation only a short period; his death having taken place on the 1st of February, 1805. Cut off in the flower of his age, he was unable to finish several important works which had cost him much labour. The result of his studies, therefore, appeared in several separate memoirs; and they relate to various subjects in physics and natural history. M. Bory de Saint-Vincent has connected his name with a branch of science to which he was much attached, by naming after him a genus of plants (*Draparnaldia*) of the family of *Algæ.*—J. D.

1. *Tableaux des Mollusques* Terrestres et Fluviatiles

de la France.   Montpellier et Paris, 1801.   A small
octavo pamphlet.

2. *Mollusques Terrestres* et Fluviatiles de la
France, Histoire Naturelle des.   Paris, 1805.   4to.
This valuable work is indispensable to every conchologist.   The arrangement, with one or two exceptions, is modern.   The animals, as well as the
shells, are described with critical accuracy, and the
plates (where nearly every species is figured) are
superior to any others we have seen.   The price,
also, places it within the reach of almost every one.
To the British conchologist it is indispensable.

DRURY, D.—*Entomology.*

An opulent jeweller of London, and one of the ardent
collectors of insects in the days of Moses Harris.   He
employed Smeathman to visit the coast of Sierra Leone,
for the purpose of collecting the insects of Western
Africa; and published an account of the most remarkable or little-known insects in his possession.   His
cabinet was sold by auction after his death, which happened about 1803.   His descriptions, in general, are
devoid of interest or scientific information; but there
are several valuable notices respecting the habits of the
Libellulidæ, and of the insects of Sierra Leone; the latter
being taken from the notes of Mr. Smeathman.   The
plates are beautifully and very correctly executed by
Moses Harris; and the early copies, which we esteem
the best, are very correctly coloured.

Illustrations of Natural History, wherein are
exhibited Figures of Exotic Insects (described in
English and French).   London, 1770—1782.   3 vols.
4to., each containing 50 coloured plates.   A new
edition has been edited by Mr. Westwood, of which
we have only seen the letterpress.   The editor's
nomenclature of the *Lepidoptera* evinces but little
acquaintance with that order of insects, although it

composes the great bulk of the species figured by
Drury. The adopted genera are the same as those of
Latreille, which are now become families ; while the
editor has passed over all those figured and defined, for
the first time, in the " Zoological Illustrations,"— a
work he only quotes when it is to be criticised. In
other respects, there are many useful observations and
much new information scattered through his volume.

DUBOIS, CHARLES. — *Conchology.*

Formerly the chief zoological auctioneer in London,
and an ardent collector of shells.

An Epitome of Lamarck's Arrangement of Testacea
or Shells, with illustrative Observations. London,
1823. 8vo. pp. 312. Very useful for young students,
who will thus understand the different genera sepa-
parated from those of Linnæus.

DUFOUR, LEON. — *Entomology.*

One of the most eminent and learned entomologists
now living. His numerous writings, unfortunately, are
scattered in the bulky volumes of the *Annales du
Muséum d'Histoire Naturelle*, in the *Annales des
Sciences Naturelles*, the *Journal de Physique*, and many
others ; so that they have become altogether inaccessible to
the great body of entomologists. How much is this to
be regretted ! According to Cuvier, they all relate to
entomological subjects of great interest, which, if collected into one volume, would be much sought after,
and universally quoted.

DUFTSCHMID, G. — *Entomology.*

One of the professors at Lintz.

Fauna Austriæ (in German). Leipsic, 1805—
1812. 2 vols. 8vo.

DUHAMEL DU MONCEAU. —*Ichthyology.*

Born at Paris in 1700, and died in 1782. He was
a physician; and his works, both on physic, agriculture,
and natural history, have attained great reputation.

Traité Général des Pêches. Paris, 1769. folio.
Enriched by a great number of excellent figures.

DUMÉRIL, CONSTANT. — *General Zoology.*

Professor of Medicine and Member of the Academy
of Sciences in Paris. He was born in 1774. His works
are in much repute.

1. *Zoologie Analytique.* Paris, 1806. 8vo. This
small volume contains an amazing quantity of information, arranged with great perspicuity.

2. *Traité Elémentaire* d'Histoire Naturelle. Paris,
1807. 2 vols. 8vo. Two editions have been published.

3. *Considérations Générales* sur les Insectes, où
l'on traite du Rang que les Insectes paraissent d'avoir
occupés dans l'Echelle des Etres; de leur Classification,
et de leur Distribution en Genre, &c. Paris, 1823.
1 vol. 8vo. with 80 plates. Although not so phy-
losophic as the above title would seem to imply, this
is a valuable little work, and the figures are remarkably good.

DUNN, ROBERT. — *Ornithology.*

The Ornithologist's Guide to the Islands of Orkney
and Shetland. We have seen only the prospectus.

DUPONCHEL, A. J. —*Entomology.*

An able entomologist of France, author of several
papers, and continuator of Godært's *Histoire Naturelle
des Lépidoptères de France,* beyond the fifth volume.

He is also the writer of a valuable monograph of the coleopterous genus *Erotyle* in the *Mémoires du Muséum d'Histoire Naturelle,* vol. xii.

EDWARDS, GEORGE. — *Ornithology.*

An unscientific but very accurate describer and painter of animals. He was librarian to the Royal Society, and enjoyed the friendship of the great sir Hans Sloane. His writings will always remain of paramount authority, from the faithfulness of his descriptions of many new birds, subsequently incorporated in the Linnæan System. He had the simplicity and piety of Izaac Walton, and may be looked upon as one of our greatest worthies.

1. *A Natural History of Birds,* and other rare and undescribed Animals. London, 1743—1763. 4 vols. 4to. containing 210 coloured plates.

2. *Gleanings of Natural History.* London. 3 vols. 4to. Forming a Supplement to the above.

EDWARDS, DR. MILNE. —*Comparative Anatomy.*

A learned physician, and eminent comparative anatomist, now settled in France, although the name would seem to denote him British. In conjunction with his friend and coadjutor, the celebrated Audouin, he has published many essays on the comparative anatomy of crustaceous insects and *Mollusca* in the *Annales des Sciences Naturelles,* as well as several others in his own name. Dr. Edwards does not appear to have been, as yet, the author of any distinct work.

EDWARDS. — *Ornithology.*

Discourse on the Emigration of British Birds. London, 1795. 8vo. pp. 64. Not having seen this work, we know not whether it is by George Edwards above-mentioned.

EISENHARDT, CHARLES WILLIAM. — *Malacology.*

Noted for a *Memoir on the Medusæ*, in the Transactions of the Academia Naturæ Curiosum of Bonn.

ELLIS, JOHN. — *Zoophytology.*

The name of Ellis will be immortalised ; for he was the first who demonstrated that corals and corallines were not plants, but animals. We regret, therefore, that so little information exists regarding many of the circumstances of his life. Sir J. E. Smith affirms that he was a native of Ireland. The date of his birth must have been about the year 1710. He is likely to have removed to London at an early age; for we find that, in his youth, he engaged there in those mercantile affairs in which he continued through life, apparently with no great success; for we find him occasionally alluding, in his correspondence, to his misfortunes and distresses. He ultimately, however, derived a comfortable income from some West Indian and American agencies, to which he was appointed by the chancellor Northington.

He early began to study natural history, and was soon connected with the Royal Society, where he gained the friendship of the first scientific men of that day. He was most attached to botany; and his connection with foreign trade afforded him excellent opportunities for introducing exotic plants, both ornamental and useful. Many of his botanical papers will be found in the Philosophical Transactions of the period : others of them, especially those of an economical nature, were published separately ; such as his Historical Account of Coffee, Description of the Mangostan and Breadfruit, Account of Venus's Fly-trap (*Dionæa muscipula*), &c.

He is said to have had his attention first drawn to the nature of sea-weeds and corallines, by observing the beauty of their ramifications when laid out on paper ;

as he was fond of amusing himself by making imitations
of landscapes, by a skilful disposition of them. On applying a microscope, he at once suspected, from their
texture, that corallines were more of an animal than a
vegetable nature. These suspicions were communicated to the Royal Society in June, 1752. He was
urged, by many of the members, to make further investigations. This he continued to do with the utmost
ardour and skill; and the result was, his entire conviction " that these apparent plants were ramified animals."

The interest which the ascertainment of this important fact imparted to the study, led him to prosecute it
with augmented zeal, both for the purpose of obviating
objections, which many continued still to urge, and of
becoming acquainted with the species. With this view,
he visited the island of Sheppy, accompanied by a
draughtsman, in the autumn of 1752. A similar excursion to the coast of Sussex was made in June, 1754,
in which he enjoyed the assistance of Ehret, who was
both a skilful botanist and artist. The fruits of this
expedition appeared in the 48th vol. of the Transactions
of the Royal Society, in the form of various letters to
the well-known naturalist, Peter Collison. Another expedition, to the north coast of Kent, was undertaken in
the autumn of the same year, along with Œder, a distinguished Danish botanist; and many important points
were determined regarding the mode of propagation
among corallines.

The result of all his investigations on the subject
was laid before the public in 1755, in his *Essay towards a Natural History of the Corallines, and other
Marine Productions of the like Kind, commonly found
on the Coasts of Great Britain and Ireland;* a work
which has established his character as an accurate and
philosophical observer, and has been the source of nearly
all our knowledge respecting these productions almost to
the present day. Whatever occurred to him subse-

quently to the publication of this work, as further illustrative of the subject, was communicated to the Royal Society, and will be found in their Transactions.

His opinions respecting the animality of corallines, although founded on such demonstrable facts, were very far from meeting with universal concurrence. Some of the principal naturalists of the day opposed them ; Dr. Job Baster, if not the most able, was one of the most violent, of his combatants ; Pallas, sir John Hill, and many others, adhered to the old opinion. Even Linnæus, his frequent correspondent, adopted his views in a very modified sense,— regarding the natural productions in question as partly of a vegetable and partly of an animal nature ; vegetables with respect to their stems, and animals with respect to their inflorescence. The term Zoophyte may be considered as embodying this idea.

Difference of opinion still prevails with regard to the nature of some few of the productions which Ellis investigated ; but, upon the whole, his labours must be regarded as constituting an epoch in the history of natural science. Although the discovery of the animality of some of the objects in question had been made on the Continent, before his time, by Peyssonnel, Jussieu, and others *, it does not appear that he was acquainted with the fact ; but, at all events, the steps he took to demonstrate it to the naturalists of this country, and his indefatigable zeal in determining and describing the species, entitle him to the highest praise. As the most effectual means of evincing the sense they entertained of his merits, the Royal Society (on the 30th of November, 1768) presented him with the Copley medal.

Various objects of interest in natural history occasionally engaged his attention, besides those branches which may be considered as more peculiarly his own province. Thus, for example, we owe to him an account of the *Siren Lacerta* of Linnæus, a treatise on the windpipes of birds, &c.

* On this subject, the history of Zoophytology in Dr. Johnston's " British Zoophytes," may be consulted with advantage.

N

His death took place on the 15th of October, 1776.
He left an only daughter, Martha, who became the
wife of Alexander Watt, Esq., of Northaw, Herts.
The greater part of his museum was sold by auction in
London, June, 1791. Linnæus has commemorated his
learned correspondent, by naming after him a genus of
plants, *Ellisia.*— J. D.

ENGRAMELLE, MARIE DOM. JOSEPH. — *Entomology.*

A monk of the order of St. Augustin.   Born at Paris
(according to Cuvier), in 1727 ; died in 1780.

Papillons d'Europe, peints d'après Nature par Er-
nest, et décrits par le Révérend Père Engramelle.
Paris, 1779—1793. 8 vols. 4to. The colouring of the
plates is very faithful, although the figures are drawn
in bad taste. Cuvier says that there are 6 volumes
in small folio, containing, in the whole, 342 plates.
The only edition we have seen is in 8 volumes, 4to.,
which generally ends with Plate 342. : a subsequent
number, however, was published, containing an addition of 16 plates ; but this, from some cause or
other, is now very rare.

ERXLEBEN, C. P. — *Mammalogy.*

Systema Regni Animalis.   Classis 1. Mammalia.
Lips. 1777. 8vo. pp. 636.

ESPER, E. I. C.— *Entomology,* &c.

1. *Die Schmitterrlingen* in Abbildungen, &c. The
Lepidopterous Insects of Europe, figured and described from Nature (in German). Erlang. 1777—
1794. 5 vols. 4to.
2. *Pflanzenthiere,* &c. The Zoophytes, figured
and described from Nature (in German). Nuremb.
1791—1800. 4 vols. 4to. The plates of both these

works are tolerably correct ; but, like most of those
executed by the artists of Germany, they are very deficient in taste and beauty.

FABRICIUS, OTHO. — *Zoology.*

A clergyman long resident in Greenland. The great
accuracy of his descriptions renders them highly valuable. No one has investigated the inhospitable shores
of that country with so much diligence and success.

Fauna Groenlandica, &c. Copen. et Leip. 1790.
1 vol. 8vo.

FABRICIUS, J. CH. — *Entomology.*

The writings of this celebrated entomologist desig-
nate that epoch in the science he cultivated, which inter-
vened between Linnæus and Latreille ; and he equally
shares with them the honour of bringing it to its present state. He was born in the duchy of Sleswick,
in 1742 ; and from thus being not very distant from
Sweden, he availed himself of every opportunity to
visit Upsal, and profit by the instructions of the great
Swedish naturalist. Linnæus, on the other hand, regarded him with peculiar favour, and their friendship,
alike honourable to the master and the pupil, remained
uninterrupted until severed by death. Fabricius was
educated for the medical profession ; and, having gone
through the usual course of preliminary studies, took
out his diploma as physician at the age of twenty-five.
But his early love for the study of nature was so much
increased by his intercourse with Linnæus, that entomology became his predominant passion. It has been
well observed by Latreille, whose biographic eloge of
this great man has furnished the materials for this
sketch, that no department of zoology stood so much in
need of revision, as entomology. Linnæus, from the
vastness of the objects over which he spread his mind,
had only seized on the prominent groups of insects ;

leaving to others the more laborious task of marshalling
the contents of each into systematic order. The first
writer who commenced this improvement was Geoffroy,
who soon after wrote on the insects found in the environs of Paris, in which he was assisted by Fourcroy;
but Fabricius resolved to prosecute the subject on a
much more extensive scale. The classification of insects,
about this time, was founded, essentially, upon one or
other of the following principles: — Those who took
Swammerdam for their guide, considered that metamorphosis should form the basis of every arrangement;
and that the transformations these animals went through
were, of all other characters, the most important.
Others, adhering to the opinions of the ancients, which
had been adopted by Linnæus, founded their arrangement on the organs of motion; justly considering that
characters drawn from the perfect insect, were much
more important than those which existed only in its im-
mature state. Dissatisfied with the exclusive attention
then bestowed to these points only, Fabricius conceived
the plan of forming a new method of classification,
chiefly founded on the structure of the mouth. This
idea was not, indeed, perfectly original,— for it had been
acted upon with signal success in the arrangement of
the vertebrated animals, and both Réaumur and Scopoli
had paid much attention to these organs: but Fabricius
pushed his examination much further; and, after many
years' attention to the subject, he gave the results to the
world in his *Systema Entomologiæ;* which was soon after
followed by the *Genera Insectorum,* wherein the defi-
nitions of many new groups, uncharacterised by his
predecessors, are enumerated with great precision. As he
advanced further in this novel method of arrangement,
he thought it expedient to follow the example, given by
Linnæus in botany, of concentrating the axioms of entomology in a separate work; and this led to the publication of the *Philosophica Entomologica,* a thin volume,
which appeared in 1778. From that period until his
death, which comprised an interval of near thirty years,

he laboured incessantly upon this basis of classification ; improving, in every succeeding work, the characters of his genera and species, augmenting both, and prodigiously extending the boundaries of his favourite science.

If Latreille is correct in the date, it would appear that Fabricius was invested with the professorship of natural history and rural economy, by his patron, the prince royal of Denmark, when only in his twenty-third year. This, of course, induced him to abandon all thoughts of pursuing the profession of medicine, and gave him an official rank in science seldom attained at so early an age. In gratitude to his royal patron, he soon after composed and published the biography of his father, Frederic IV., one of the most excellent sovereigns of Denmark. Nor was this the only occasion on which he proved himself something more than a mere entomologist. Intimately acquainted with the dead and living languages, he travelled over the northern, and in most of the central, states of Europe ; not merely for scientific information, but to observe the manners, customs, laws, and institutions of each. The results of nearly all these excursions he gave to the world, although we have to regret they are but little known in this country. Latreille mentions, that his travels in Norway (originally published, in all probability, in the Danish language) was translated into French by Millin; and that this was followed by his excursion to Petersburgh, chiefly undertaken with a view to an examination of the waters of the Neva, the use of which, at that time, was considered as hurtful to foreigners as it was beneficial to natives. Most of all do we regret that his travels in England should not have been translated into our language; since it probably contains many interesting anecdotes of the great naturalists of that period ; setting aside the interest which must ever attach to the opinions and remarks of so eminent a man on our national peculiarities. In proof of the high esteem and personal friendship with which he was regarded in this country, it is remarkable that he made no less than seven different visits to England, where he

N 3

was always received as a welcome guest by the illustrious
Sir J. Banks and Dr. Hunter; and had free access to
all the entomological treasures contained in the splendid
cabinets of Drury, Francillon, Jones, and Haworth: these
valuable additions to science have all been commemorated
in his various works; but we had never any conception,
until reading Latreille's memoir, that he had also contributed to our literature. It appears, however, that the
account of the kingdom of Denmark, published by Pin-
kerton, in his valuable geography, was actually written by
Fabricius. No one, indeed, was more qualified for the
task; for he held various civil appointments, and enjoyed
the friendship of the first men in Copenhagen. When,
or to whom, he was married, does not appear; but, for
some unexplained cause, it is said that his wife and
eldest son resided in Paris, a city he himself often visited,
and where, in such eminent men as Desfontaine, Bosc,
Olivier, &c., he possessed the warmest friends.

Being now generally admitted as the first entomologist
of his age, the private and public cabinets of every city
he visited were thrown open to him, and his extraordinary zeal was manifested in the use he made of them.
No wonder then, as Latreille observes, that he made us
acquainted with six times the number of insects enumerated in the works of Linnæus. A serious evil, however,
inevitably sprung from this plan of proceeding. Every
naturalist is aware how slight are the shades which separate species; and that it is utterly impossible for a
writer, engaged in describing species, to remember every
one, of every order, he may have previously incorporated
in his work. Unable, therefore, to revise his materials,
scattered in the different cabinets of Europe, Fabricius
was sometimes led into those errors which such a power
would have prevented. He described, as new, species
he had introduced in his previous works; and confounded
others, really distinct, with such as were well known.
But where is that naturalist, as Latreille well observes,
whose first writings, nay, his most matured, are without
errors or imperfections? Such an observation would only

have originated in a generous mind; while it is the characteristic of a little—or rather of a contemptible—one,
to single out these specks, strive to give them a prominent
position, and remain perfectly silent on all the great ex-
cellencies by which they may be surrounded. The spirit
of analysis and of rigid comparison did not then exist,
as it does now; and naturalists were accustomed to regard as varieties, those modifications of colour or form
which are now known to indicate permanent differences.

Fabricius was no less great in talents than in mind;
— far from being jealous of his cotemporaries, he did
every thing in his power to aid their researches. No
sooner had the admirable work of Walckenaer, on the
spiders, reached his hands, than he sent him, unasked,
the whole of his collection of those insects for his examination: the same liberal spirit influenced him in
giving every assistance, towards the close of his life, to
Clairville and Spinola, whose works are now so valuable,
and who ever regarded him with respect and affection.

We have already seen that Fabricius by no means
confined his studies to entomology: his mind took a
wider range; and, during his abode at Freyberg, both
mineralogy and botany engaged much of his attention.
In pursuit of the latter, he was constantly seen with his
scholars roving through the environs of Kiel, gathering
the native plants, and forgetting his own fatigues in the
pleasure he felt in imparting instruction to others. Latreille even mentions that he published some works, in
German, both on botany and agriculture; but we are
ignorant of their titles. His kindness to his pupils is
said to have amounted almost to parental affection; and
to one of them, M. Weber, he dedicated a small work,
containing an abridgment of his lectures on natural history. Those high moral qualities, which thus endeared
him to all his private friends, led him, also, to take a
lively interest in every thing which concerned the public
good and the happiness of his fellow creatures. He
warmly interested himself, as Latreille observes, in the
fate of those unfortunate children who were abandoned

to public commiseration.*   To forward this object, he
wrote treatises on hospitals and public institutions,
as well as the civil and medical regulations by which
they should be governed : these excited a considerable
sensation in Germany, and produced some beneficial
changes.   They should, observes Latreille, have long
ago been translated into French ; but the pen, he continues, of the greater part of our translators, is too exclusively employed upon works, the reading of which
produces those baneful eyils which Fabricius, though
he strove to cure, yet hoped, at least, to diminish.

The many and expensive travels, joined to the formation of a large museum of natural history, considerably
impaired his fortune ; yet he not only refrained from
soliciting any other appointment or recompence from
his government, but seems to have refused offers which
would have led him into the service of others.   Latreille
remarks, that " Colonel" Cathcart was commissioned to
make him an offer of 200*l.* per annum for entering into
the service of the East India Company, and proceeding
with the embassy then about to sail for China.   Nothing
could have been more tempting to an entomologist than
such an appointment, honourable to himself, and promising a new field for his researches.   But the patriotic
feelings of Fabricius would not allow of this separation
from his native land, and he at once declined the offer.   His
disposition, indeed, seems to have been a singular union
of energy and modesty — of simplicity and patriarchal
goodness.   With all his talents, he carried these feelings
so far, that his friends reproached him with his too great
diffidence.   Latreille speaks of him with the affection of
an attached brother.   Amid the smothered enmity, and
the secret slander, so often manifested by the followers of
science towards each other, it is delightful to see such
opposite feelings uniting two men, pursuing the same
track, and equally ambitious of fame.

The political changes and misfortunes of his native

* This passage in M. Latreille's memoir is somewhat obscure ; it would
seem to indicate those " unfortunate" children who were foundlings.

country, towards the close of his life, seem to have had
a very unusual effect, both upon his health and constitution; for, according to the belief of his widow, they
shortened his days. His vigorous constitution gave
every hope of a long life; but his friends soon perceived
that the deplorable situation of Denmark affected him
most painfully. His expressive physiognomy, which
used to be habitually serene, and even gay, became
clouded with sorrow and anxiety. His attachment,
also, to the French party, which he could not conceal,
created him many enemies among his fellow citizens.
Their slanders, however, had no effect in diminishing
his influence at court; although they seem to have
much disturbed his habitual serenity of mind. How far
these causes accelerated that disease which eventually
terminated his valuable life, does not exactly appear;
but, at the age of sixty-five, he was somewhat suddenly taken away by dropsy, leaving a widow and two
sons, with a host of ardent friends and disciples, to
lament his loss. Latreille concludes his diffuse, but
interesting memoir, by intimating he had written his
own biography; but if so, it has, unfortunately, never
been published.

The voluminous works of Fabricius, on entomology,
will ever remain established books of reference: to
these, only, will our list be confined.

1. *Systema Entomologiæ.** Flensburgi et Lipsiæ,
1775. 1 vol. 8vo.

2. *Philosophia Entomologica,* sistens Scientiæ Fundamenta. Hamburgi, 1778. 1 vol. 8vo.

3. *Species Insectorum,* exhibentes eorum Dif-
ferentias specificas, Synonyma Auctorum, &c. Hamburgi, 1781. 2 vols. 8vo.

4. *Mantissa Insectorum,* sistens eorum Species
nuper detectas. Hafniæ, 1787. 2 parts, or volumes,
8vo.

5. *Entomologia Systematica,* emendata et aucta,

* This work, by some singular oversight, is omitted in Brunet's *Manuel
du Libraire.*

cum Supplemento et Indicibus. Hafniæ, 1792—99.
In all, 9 vols. 8vo.

6. *Systema Eleutheratorum,* secundum Ordines, Genera, et Species, adjectis Synonymis, Locis, Observationibus, Descriptionibus.   Killæ, 1801. 2 vols.
8vo.

7. *Index Alphabeticus* in Systema Eleutheratorum.
Brunsw. 1 vol. 4to.

8. *Systema Rhyngotorum.*   Brunswigæ, 1803.
1 vol.— Index Alphabeticus, do.   Brunsw. 1805. in
4to.

9. *Systema Antliatorum.*   Brunsw. 1804. 1 vol.
8vo.— Index Alphabeticus, do.   Brunsw. 1805. in
4to.

10. *Systema Piezatorum.* Brunsw. 1805.— Index
Alphabeticus, do. do.

11. *Systema* Glossatorum.   Brunsw. 1806. in
8vo.*

Several other of his minor works on Entomology are
enumerated by Percheron, in his *Bibliographie Entomologique,* i. 105.

FALLEN, CHARLES FRED. — *Entomology.*

An acute entomologist, and Professor of Natural History at Lund. He seems to have more especially devoted
his attention to the dipterous insects of his native
country.

Diptera Sueciæ. Lundæ, 1814—1817. 1 vol. 4to.
We know not whether any other has yet appeared.

FAUJAS DE ST. FOND, B.

Professor of Geology in the French Museum of Paris
(*Muséum d'Histoire Naturelle*).

Histoire Naturelle de la Montaigne de St. Pierre

---

* This title is copied from Brunet's *Manuel.* I rather think, however,
that this volume was never published. Illiger has given the characters of
the genera, and these have been copied in several more recent publications.

de Maestricht. Paris, 1799. 1 vol. 4to. Many interesting fossil animals are mentioned in this work.

FAVANNE, M. DE. — *Conchology.*

A French conchologist of slight merit. He is mentioned as the author of a *Dictionnaire de Conchologie*, which we have never met with : he also edited the second edition of Argenville's Conchology; which see.

FERMEN, P. — *General Zoology.*

A Dutch naturalist, long resident as a physician at Surinam. Many of his observations are worth consulting, but the nomenclature is very erroneous.

1. *Histoire Naturelle de la Hollande Equinoxiale.* Amsterdam, 1765. 8vo.

2. *Description de Surinam.* Amst. 1769. 2 vols. 8vo.

FÉRUSSAC, J. D'AUDEBERT DE. — *Conchology.*

An eminent French conchologist, who particularly studied the land shells.

1. *Essai d'une Méthode Conchyliologique.* Paris, 1807. Pamphlet in 8vo. This we have never seen.

2. *Mollusques Terrestres et Fluviatiles,* Histoire Naturelle, générale et particulière, des. Folio. The execution of this work is magnificent; but its enormous price precludes it from being of general utility. A synoptical abridgment would be of much use. We believe it has not yet been completed.

FICHTEL, LEOPOLD DE. — *Conchology.*

This naturalist, attached to the Vienna Museum, and J. P. C. de Moll (of the Royal Academy at Munich), published conjointly —

Testacea Microscopica aliaque minuta ex Generibus Argonauta et Nautilus. Vien. 1798. 4to. pp. 123. pl. 24. The text is in Latin and German. A subsequent edition was printed at Vienna.

FISCHER DE WALDHEIM.—*General Zoology.*

An eminent and learned zoologist, director of the Imperial Museum at Moscow. Most of his writings are in German, and hence, unfortunately, are little known in this country. He is at present engaged on a monograph of the birds belonging to the genus *Carbo.*

1. *Fragments* of Natural History. Francfort, 1801. 1 vol. 4to.

2. *Anatomy* of the Makis. Francfort, 1804.

3. *Description of several new Insects.* Originally inserted in the Moscow Transactions, but afterwards printed separately. Moscow, 1806. 1 vol. 4to.

4. *Entomographica* Imperii Russici. Moscow, 1820—22. 2 vols. 4to. "with splendid engravings." This we have not seen.

5. *Observations on the Medeterus,* a Carnivorous Dipterous Insect. Moscow, 1819. 4to. with plates.

6. *On the Argas* of Persia. Ib. 1823. 4to. with a plate.

7. *On the Physodactylus,* a new Genus of Coleoptera. Moscow, 1824. 8vo.

8. *Notice d'un Animal Fossile* de Sibérie, inconnu aux Naturalistes. Moscow, 1808. 4to. 2 plates.

9. *Notices des Fossiles* du Gouvernement de Moscou. Ib. 1809. 4to. 3 plates.

10. *Synopsis Mammalium.* Lipsiæ, 1830. 4to. We find this work mentioned in a London Catalogue, but have never seen it.

FISCHER, GOTTHELD. — *Zoology.*

This zealous naturalist, a native of Germany, is Professor of Natural History at Moscow.

1. *Fragments* of Natural History (in German).
1801. 1 vol. 4to.
2. *Anatomy of the Makis* (in German). Francfort, 1804.

FITZINGER. — *Erpetology.*

A physician of Vienna, and one of the most distinguished erpetologists of Germany.

New Classification of Reptiles, according to their
natural Affinities (in German). 1826. 1 vol. 4to.
This work we unfortunately do not possess, and the
foundations of the author's natural system are therefore unknown to us. His divisions, however, as noticed in other works, are excessively numerous, and
do not appear to be regulated by any comprehensive
or fixed principles.

FLEMING, THE REV. JOHN. — *General Zoology.*

A clergyman of the Scotch kirk, who has given a
useful and well digested compilation from Cuvier, &c. on
the internal structure of animals, under the title of

1. *Philosophy of Zoology.* Edinburgh, 1822. 2 vols.
8vo.
2. *Natural History of British Animals.* 1 vol. 8vo.
Although numerous errors will be found in this work,
there are some good and original observations scattered through it.

FORSKAHL, P. — *Zoology and Botany.*

Peter Forskahl was an intelligent traveller and accomplished naturalist, one of the many whose exertions
in behalf of science have been the means of bringing
them to an early grave. He was a Swede, born in
1736. When very young, he was sent to prosecute his
studies at Gottingen, where he distinguished himself

beyond most of his companions by his application and proficiency. After completing his *curriculum*, he returned to Sweden, and, among other pursuits, followed that of natural history. This naturally led him to court the society and friendship of Linnæus, whose favour he soon gained. Linnæus alludes to him, in one of his letters, as his excellent pupil, from whom many interesting discoveries might be expected; and as particularly excelling in the knowledge of insects, although very well versed in the other branches of natural history.

Frederick I. king of Denmark, having resolved to send out a scientific expedition to examine certain parts of Asia, particularly those countries to which such an intense interest attached as having been the scene of the events recorded in Scripture, Forskahl, on the recommendation of Linnæus, was chosen one of the members. The others were Niebuhr, Cramer, Carsten, Von Haven, and Baurnfeind. Forskahl was exceedingly well qualified for the task entrusted to him; for, besides his knowledge of natural history, he had an extensive and accurate acquaintance with Oriental languages. But his career of usefulness was destined to be brief; for he was seized with the plague at Jerim, in Arabia, and died on the 11th of July, 1763. The valuable materials he left behind him prove how zealously he had laboured, after reaching the countries he was sent out to examine. They were prepared for the press by Niebuhr, the only one of the party that survived; and he executed the task so faithfully, that the Stockholm Academy of Sciences elected him a member in token of their approbation. These posthumous works (the particular titles of which are given below) contain a systematic catalogue and description of about 300 animals of various classes, with the names in Latin, Greek, and Arabic; a list of the chief medicines found in the great laboratory in Cairo; a *Flora Ægyptiaco-Arabica;* and forty-three plates (with two leaves of letterpress), twenty of which represent plants, and twenty-three animals. Those works were highly valued by Professor Michaëlis, of Gottingen, one of the best Oriental scholars of his day, as illustrative of the text both of the Old and New Testament, in those places where allusion is made to natural objects. No one more deeply lamented than he, the premature decease of Forskahl; for his powers of observation, assiduity, and earnest desire to make every sacrifice to establish the truth, were qualities which are not very frequently to be met with in travellers.—J. D.

1. *Descriptiones Animalium*, &c. quæ in Itinere Orientali observavit. Copenh. 1775. 1 vol. 4to.

2. *Icones* Rerum Naturalium quas in Itinere Orientali depingi curavit. Copenh. 1776. 1 vol. 4to.

FORSTER, JOHANN REINHOLD.—*Zoology and Botany.*

This erudite and philosophic naturalist was born at Dirschaw, in 1729. He received an offer from our government to accompany captain Cook on his second voyage, which he at once accepted; and he returned again to England, stored with an accession of knowledge in various branches of science. He subsequently, however, retired to the Continent, and was chosen Professor of Botany at Halle. This situation he held at the time of his death, which took place in 1798. His son George accompanied him round the world, and subsequently became a privy counsellor and librarian at Mayence. The zoological works of both are equally valuable with their botanical publications.

1. *Enchyridion* Historiæ Naturali inserviens. Halle, 1788. 1 vol. 8vo.

2. *Zoologicæ Indicæ* rarioris Spicilegium. London, 1790. 1 vol. 4to.

3. *Novæ Species* Insectorum, Centuria 1. London, 1771. 1 vol. 8vo.

FORSTER. — *Ornithology.*

1. *Observations* on the Brumal Retreat of the Swallow; with a copious reference to passages relating to this subject in different authors.   London. 8vo.

2. *Synoptical Catalogue* of British Birds.   London. 8vo.   We know not whether these two tracts are by either of the last-mentioned authors.

FORTIS, ALBERTO.

Librarian at the College of Bolognia.   Born at Vicenza, 1740; died in 1803.

Mémoires pour servir à l'Histoire Naturelle, et principalement à l'Oristographie de l'Italie.   Paris, 1802.   2 vols. 8vo.

FOURCROY, ANTOINE-FRANCOIS DE. — *Entomology.*

Professor of Chemistry, and Member of the Academy of Sciences in Paris.   Born in 1755, died in 1809.   At an early age he published —

Entomologia Parisiensis.   Paris, 1785. 2 vols. 8vo.

FREMINVILLE, BARON DE. — *Zoology.*

A distinguished officer of the French navy, and, as Cuvier remarks, " an able naturalist."   He is the author of various articles in the " Dictionnaire Classique d'Histoire Naturelle; but does not appear to have published any distinct work.

FREYCINET, LOUIS DE (*Voyage of*).

We have sometimes quoted the labours of the naturalists who accompanied this expedition, by the name of its commander, although he had not, himself, any hand

in their publication.   The zoological portion is written
by MM. Quoy and Gaimard.   (See Quoy.)

Fries, E. — *Botany.*

A celebrated German botanist; immortal from his
discovery of the circular theory of affinity in the vegetable kingdom, and of the true quinary system of representation therein.

Systema Mycologicum, sistens Fungorum Ordines,
Genera, et Species huc usque cognitos.   Lundæ,
1821—1823.   3 vols. 8vo.

Frisch, Jean L. — *Ornithology and Entomology.*

Director of the Gymnasium at Berlin.   Born in 1666;
died in 1743.

1. *Vorstellung der Voegel* Deutschlandes, &c., or Representations of the Birds of Germany, and of several
Exotic Species (in German).   Berlin, 1739—1763.
2 vols. folio.   The plates are 255 in number, and
very exact.

2. *Insects of Germany,* Description of the (in German).   Berlin, 1730. 4to.

Garden, Dr. Alexander.

A Scotch physician, who resided in Carolina.   He is
merely known as having been an intimate correspondent
of Linnæus, who often quotes his name.

Gaymard and Quoy. — *General Zoology.*

We quote the names of these two eminent zoologists
together, as they were the official naturalists and fellow
voyagers on two of the French expeditions sent, of late
years, to the southern hemisphere, and their discoveries
are published together.   M. Quoy has more especially

taken the department of the *Mollusques*, and has treated
their history with the hand of a master.   His discoveries
have been numerous and important.

1. *Voyage de l'Uranie*, commanded by M. Freycinet.
The zoology is contained in one 4to. volume of text,
and one folio of plates.

2. *Voyage de l'Astrolabe*. The plates printed uniformly with the above, but the text in 8vo. Is now
in course of publication, if not completed.

GEER, CHARLES BARON DE.— *Entomologist.*

The high reputation of baron de Geer, as a philosophical naturalist, rests almost entirely on his *Mémoires*
upon insects, a work which has always been regarded
as one of the most valuable that ever appeared on that
class of animals: it is full of original and important
observations, accurate descriptions, and delineations of
external structure.

Charles de Geer, baron of Leutsta, marshal of the
court of Sweden, knight of the Polar Star, and commander of the Order of Vasa, belonged to an opulent
Dutch family, which had established themselves in
Sweden in the time of Gustavus Adolphus.   He was
born in 1720 ; and, when about four years of age, accompanied his parents to Holland, where he resided till
his eighteenth year. His studies were commenced at
Utrecht, and subsequently carried on at Upsal, where
he enjoyed the combined instruction of Linnæus, Celsius,
and Klengenstiern.   At an early age, he came into possession of one of the largest fortunes in Sweden, which
in part descended to him in a direct line from Louis de
Geer, one of his ancestors, in whose person the family
was first ennobled. This individual had acquired great
wealth and reputation by the important improvements
he effected in various branches of art, particularly in
the manufacture of cast iron, and brass. He introduced
foreign artisans into the canton of Dannemora, where
his property chiefly lay ; and rose to such influence, that,

when the country was once threatened by a foreign ene-
my, he equipped a considerable fleet for defending the
coasts, entirely at his own expense.

De Geer is said to have acquired a taste for studying
the habits and structure of insects, by observing the interesting proceedings of some silkworms, which had
been given him, while yet a boy, as an amusement.
This predilection was further strengthened by his intercourse with the celebrated Muschenbroeck, and fully
confirmed by the perusal of Réaumur's admirable memoirs. But while prosecuting this pursuit, he did not
neglect others of higher national importance ; he expended large sums in improving the mode of working
the iron mines of Sweden, in repairing churches, found-
ing hospitals and schools, and promoting other philan-
thropic objects.

We do not possess any detailed account of De Geer's
life; and it was not, probably, marked by any incident
apart from his history as a man of science, that would
be of much interest to a reader of the present day. He
was a Member of the Academy of Stockholm, regularly
attended its meetings, and took an active part in its
proceedings. The first of his memoirs on insects was
published in the Transactions of that body. But in 1752
appeared the first volume of his great work, in quarto,
which, as it was suggested by Réaumur's, also assumed
its title : it contains 37 plates, and is dedicated to the
queen of Sweden. It is chiefly occupied with the
history of the various kinds of caterpillars. The second
volume did not make its appearance till about nineteen
years afterwards ; and four others succeeded it at not
very long intervals. The seventh, or last volume, was
not published till after his death. It is adorned with
an elegant portrait, but without the appropriate accom-
paniment of a biographical memoir.

De Geer was long a martyr to gout—a disorder which
terminated his honourable career on the 8th of March,
1778. His extensive museum was presented, by his
widow, to the Academy of Stockholm, where a marble

bust recalls the memory of their associate and benefactor.

There can only be one opinion of De Geer's merits.
His masterly work is a complete repertory of original
observations, accurate descriptions, and correct delineations. Professedly in imitation of Réaumur's work, it
forms an indispensable sequel to it, and supplies an immense deal of what that author left untouched. Yet,
with this similarity of purpose, there is a considerable
contrast between the two. With powers of observation,
perhaps superior to those of his celebrated cotemporary,
De Geer falls greatly short of him in the felicity of his
illustrations, and the means of interesting his readers.
The one is diffuse, redundant, and popular; the other,
concise, direct, and explicit. Réaumur is best fitted to
attract the regard of those not conversant with the subject; De Geer, to satisfy the wants of an entomologist.
De Geer's knowledge of system was as superior to
Réaumur's, as that of Linnæus was to Buffon's : in other
words, the one had no sense of its value; the other regarded it as of paramount importance. Notwithstanding the multitude of modern entomological works, we
often find that we are obliged to refer to De Geer, for
that information in which they are deficient. The facts,
of which his work is so copious a storehouse, have
scarcely yet been entirely transfused into our present
publications. This may be partly owing to the rarity
of the work. The first volume is much scarcer than the
rest, and the alleged cause is rather a curious one. It is
said, that De Geer was so ill pleased with the success of
his work, when it first appeared, that he even allowed
this feeling of disappointment to get the better of his
prudence, by committing a great part of the impression to
the flames. Out of gratitude to the purchasers of the
first volume, he is likewise said to have sent them, as a
present, a copy of each of the successive volumes as they
appeared.

De Geer was a nobleman by birth, no less than by his
splendid talents ; hence he enjoyed the favour of his

sovereign, and the respect of the learned. He did not examine nature through the distorting glass of system ; every page of his work, therefore, offers some authentic and important fact, which succeeding naturalists will derive instruction from perusing.— W. S.   J. D.

Mémoires pour servir à l'Histoire Naturelle des Insectes. Stockholm, 1752—1771. 7 vols. 4to. with plates : now become very rare. There is a German translation, nearly as scarce as the original work.

GEOFFROY, ET. LOUIS. — *Entomology and Conchology.*

A celebrated physician of Paris.

1. *Histoire abrégée des Insectes* qui se trouvent aux Environs de Paris. Paris, 1762—4. 2 vols. 4to.

2. *Traité* sommaire des Coquilles, tant Fluviatiles que Terrestres, qui se trouvent aux Environs de Paris. Paris, 1767. 1 vol. 12mo. Cuvier observes that this little work is remarkable as being an attempt to class the shells according to their animals.

GEOFFROY ST. HILAIRE, ETIENNE. — *General Zoology.*

At present one of the Professors in the Museum of Natural History, and Member of the Academy of Sciences in Paris. The numerous essays of this acute and learned man are scattered in the *Magazin Encyclopédique, Les Annales du Muséum,* and in many other periodicals : he likewise contributed much to the great work on Egypt.— W. S.

Geoffroy St. Hilaire was born at Etampes, on the 15th of April, 1772. He was educated in the College of Navarre, where he enjoyed the instructions of Brisson, who was Professor of Experimental Physics in that institution. He acquired a taste for the sciences in general under that great master : but natural history, properly so called, occupied the principal place in his regard. Mineralogy was the branch which first fixed his attention ; and in this he had the advantage to receive lessons from the celebrated Haüy; which, however, were suddenly interrupted by the eventful occurrences of August, 1792. Haüy and most of the other professors at Navarre were thrown into prison; and it was almost entirely owing to the exertions of young St. Hilaire that they succeeded in making their escape before the fatal days of the 2d and 3d of September. This circumstance had a considerable influence on his future fortunes, for it procured him friends and patrons who lost no opportunity in promoting his interests. An appeal made by Haüy to Daubenton, in favour of his young liberator, procured him the situation of sub-curator and demonstrator to the Cabinet of Natural History, when he was only twenty-one years of age. When the National Convention, in June, 1793, organised the Jardin des Plantes, he was appointed one of the twelve naturalists attached to the establishment; the vertebrate animals being assigned as his department. He was unwilling to teach any other branch of natural history than mineralogy, in which he was most versant; but yielded to the remonstrances of Daubenton, and sent for a young man from Normandy to aid him in his new labours. This was no other than Cuvier, then unknown to the world of science.

M. Geoffroy St. Hilaire visited Egypt in 1798, ascending as far as the cataracts, assisting in establishing an Institute of the Arts and Sciences at Cairo, and passing three entire weeks in the catacombs of Thebes. He was nominated a member of the Institute of France on the 14th of September, 1807; and Professor to the Faculty of Sciences on the 20th of July, 1809. In 1808, he was entrusted with a mission to Portugal, in which he was exposed to many dangers, having been imprisoned at Merida, and narrowly escaping with his life. He joined the French army for a time during the campaign which terminated in the battle of Vimiera.

In 1815, he became a member of the chamber of deputies for the town of Etampes. He is still alive, and in the enjoyment of that respect which his eminent talents so justly entitle him to. — J. D.

In conjunction with Frederick Cuvier, he published, in numbers —

   1. *Histoire Naturelle des Mammifères.* Paris, 1819. Nos. 1. to 22.: each contains 6 coloured plates.
   2. *Philosophie Anatomique.* Paris, 1818. 2 vols. 4to.

GEOFFROY ST. HILAIRE, DR. ISIDORE. — *Zoology.*

One of the principal officers of the Muséum d'Histoire Naturelle, and son of the above. A zoologist of high talent, assiduous application, and engaging manners. He is the author of numerous memoirs in the French Transactions, of the greater part of the ichthyology, of the large work on Egypt, and is more or less concerned in almost every zoological work of repute published in Paris. Science, we trust, will be benefited, ere long, by a distinct volume from this talented naturalist.

GERMAR, ERNEST FREDERIC. — *Entomology.*

An able and industrious entomologist, and Professor of Mineralogy at Halle.

   1. *Dissertatio* sistens Bombycum Species, &c. Halæ, 4to.
   2. *Magazin der Entomologie.* Halle, 1813—1821.
   3. *Insectorum Species Novæ* aut minùs cognitæ, Descriptionibus illustratæ. Vol. I. Coleoptera. Halæ, 1834. 8vo.
   4. *Systematis* Glossatorum Prodromus, systens Bombycum Species secundum Oris Partium Diversitas in Nova Genera, distributas. 2 parts. 4to. Lipsiæ, 1810—1812.

GESNER, CONRAD. — *General Zoology.*

One of the ancient fathers of zoology; whose laborious work on animals exhibits, in itself, a perfect picture

of the then state of zoological knowledge. Although a
compilation, Cuvier observes, it is enriched with many
useful observations. The woodcuts are numerous, and
many of them very good.

1. *Historia Animalium.* 4 vols. folio. Tiguri,
1551—58.

2. *Vögel-Buch,* Thierbuch und Fischbuch, teutsch,
durch R. Heuszlin und C. Fover. Numerous woodcuts, neatly coloured. Zurich, 1653—75—82. 1
thick folio. (*Bohn's Catalogue.*) This would seem
a partial translation of the above.

GILLIAMS.—*Zoology.*

An American naturalist, but little known in this
country. According to Cuvier, he is the author of
several memoirs on reptiles and fishes, published in
the " Journal of the Academy of Natural Sciences of
Philadelphia."

GIOENI, GUISEPPE.—*Conchology.*

A Sicilian writer, remarkable for having committed
the mistake of supposing the testaceous stomach of the
Bulla lignaria Linn. to be a new genus of shells !

Descritzione, &c. ; Description of a new Family
and a new Genus of Testacea (in Italian). Naples,
1783. 8vo.

GMELIN, JOHN FRED.

Born at Tubingen, in 1748. Professor of Chemistry
at Gotfingen ; and the great compiler of the last edition
of the Systema Naturæ of Linnæus.

Caroli a Linné, Systema Naturæ per Regna Tria
Naturæ. Editio decima tertia, aucta, reformata.
Curâ J. Fred. Gmelin. Lipsiæ, 1788. 3 vols. (in ten
parts) 8vo.

In examining this work, we cannot help being struck

by the immense labour and unwearied industry that must
have produced it, and regretting the deficiency of judgment of the worthy editor. Cuvier very justly says it is
" tout indigesté et dénué de critique et de connaissance
des choses." Yet, as being the latest work professing
to describe all the known species, it is, in some measure,
of use. Dr. Turton has given a literal translation of
the whole, in 7 vols. 8vo. This, for the above reasons,
is worthy of consultation.

GMELIN, SAMUEL THEOPHILUS.* — *Zoological Traveller.*

Born at Tubingen, in 1743. An enterprising traveller and acute naturalist, employed in the service of
Russia from 1768 up to the period of his death, which
happened in Persia, in the year 1774.

His *Travels*, which we have not seen, were published
in German, at St. Petersburg, in 4 vols. 4to. 1770—
1784; and are stated by Cuvier to abound in valuable
notices on natural history. He seems to have been succeeded, in the same capacity, by Guldenstedt.

Reisen durch Russland, &c. Travels through
Russia (in German). Petersburg, 1770—1789.
4 vols. 4to.

GODART, J. B. — *Entomology.*

One of the best entomologists, and most accurate de-
scribers, of his age. He was Chief Professor in the
Lyceum of Bonne, and died in 1825. His admirable
account of the diurnal butterflies is contained in the
entomological portion of the

1. *Encyclopédie Méthodique*, where it forms the
article *Papillon.* This invaluable treatise is indispen-

---

* There seems to me some confusion about these Gmelins. Brunet calls
one, *Sam. Gottl.*, who appears to be the same as Cuvier's *Sam. Theophilus.*
He mentions another (*J. Georg.*) as having published Travels in Siberia,
from 1733 to 1743, Gotting. 1751—2, in 4 vols. 8vo. This latter name
is not found in Cuvier's list. Nor is that of *John Fred.* Gmelin noticed by
Brunet. (See *Manuel*, ii. 36.)

sable to all who study this charming family; for it
has never yet been supplanted, although published
in 1819.

2. *Histoire Naturelle* des Lépidoptères ou Papillons de France; commenced in 1822.   He lived
only to complete the first five volumes.

GOEDART, J. — *Entomology.*

John Goedart (or Goedartius, as it may be convenient to call him, to distinguish him from a modern
French naturalist of high merit) was a Dutch painter;
and distinguished himself by investigating the metamorphoses of insects, and delineating them with considerable accuracy, at a time when little attention was
bestowed on such subjects.   He was born in 1620, and
died in 1668.   We are not aware that any records
exist of his life; the incidents of which, however, cannot
now be supposed to be of much interest to the reader.
He probably, at first, regarded insects merely as beautiful objects for exercising his pencil, as has been the
case with many painter-entomologists; and from con-
templating their forms, became gradually to be interested
in their history.   The work in which his researches
were laid before the public, was written in Dutch, and
published at Middleburg in 1662, 3 parts, 8vo. with
500 coloured plates.   A Latin translation soon appeared,
of which the title was, " Metamorphoses et Hist. Nat.
Insectorum, cum Commentario Jo. de Mey, et duplici
ejusd. Appendice, una de Hemerobiis, altera de Naturâ
Cometarum. (!)   Middleburgh, 1662—1667."   The
second volume of this edition contains a memoir, by
Paul Voezaerdt, on the origin and utility of insects.
The work is best known in this country by Dr. Martin
Lister's editions; one of which, in English, with notes,
was printed at York, in 1682, 4to.; another in Latin,
very much altered, and arranged according to a method
of his own, appeared at London in 1685, 8vo.   To

this was added a new edition of Lister's Appendix to
his " Historia Animalium Angliæ," and four new plates
of beetles, &c. In addition to these, a French translation
of the work was published at Amsterdam in 1700,
3 vols. 12mo.*— J. D.

GODMAN, DR.— *General Zoology.*

American Natural History. Mastology. Philad.
1826. 2 vols. 8vo. with very good plates.

GOEZE, J. A. E.— *Intestinal Worms.*

A clergyman of Queblembourg. Born in 1731; died
in 1793.
Hist. Nat. des Vers Intestineux (in German).
Blankenbourg, 1782. 1 vol. 4to.

GOLDFUSS, G. A. — *General Zoology.*

A German zoologist, and Professor of Natural History
at Bonne.
Manual of Zoology (in German). Nuremberg, 1820.
2 vols. 8vo. He has likewise written papers in various
Transactions, and is now publishing a large work on
Petrifactions.

GOUAN, ANTOINE. — *Ichthyology.*

A learned naturalist, and professor at Montpellier.
Historia Piscium, sistens ipsorum Anatomen atque
Genera, Latine et Gallice.   Strasbourg, 1770. 4to.
pp. 252. pl. 4.   The genera are alone described, but
in considerable detail.

GRAUMANN, P. B. C.— *Mammalogy.*

Introductio in Historiam Naturalem Animalium
Mammalium.   Rostochi, 1778. 8vo. pp. 90.

* For further remarks on this author, see Preliminary Discourse, p. 20.

GRAVES, GEORGE. — *Ornithology.*

Professional colourer of natural history plates.

1. *British Ornithology.* 3 vols. royal 8vo. London.
Three or more volumes are already published, each
containing 48 plates.

2. *Ovarium Britannicum,* or Figures of British
Eggs.

GRAVENHORST, J. L. C. — *Entomology.*

An assiduous entomologist, Member of the Physical
Society of Gottingen, &c. Among his chief works are —

1. *Coleoptera Microptera* Brunsvicensia. Bruns.
1802. 1 vol. 8vo.

2. *Monographia Coleopterorum* Micropterorum.
Gottingæ, 1806. 1 vol. 8vo.

3. *Monographia* generis Ichneumonum. 1814. pl.
8vo. Vol. I.

GRAY, JOHN EDWARD. — *General Zoology.*

One of the officers of the British Museum. An
eminent erpetologist and conchologist, author of various
papers in scientific journals and other periodicals. He
has also edited or named the plates published at the
expense of general Hardwicke, on the animals of India,
and was to have written the descriptions necessary to
complete the work; but disputes arose, and it is now in
Chancery. Mr. Gray has also commenced several technical scientific works in which new species are described;
but we are unacquainted with any one which has yet
been completed; and many others have only been announced. He has assisted Mr. Griffith in the supplementary species of birds added to his translation of
Cuvier, and has more recently edited an edition of
Turton's "Land and Fluviatile Shells," to which he
has made considerable additions.

GRAY, GEORGE. — *General Zoology.*

Brother of the foregoing, and also an officer of the
British Museum. An able entomologist, who has begun
a valuable work, entitled—

　1. *Entomology of Australia,* in a Series of Monographs. Part I. the Genus Phasma. 4to. Lond.
1833. With 8 admirable plates by Mr. Ch. Curtis.

　2. *Synopsis of the Phasmidæ.* 8vo. pamph. London, 1835.

Mr. Geo. Gray has likewise written several papers
in the "Transactions of the Entomological Society of
London," &c.

GREW, N.— *Botanist and Zoologist.*

Secretary to the Royal Society. He is celebrated for
his discoveries in vegetable physiology; but his zoological works are few and commonplace. He died in
1711.

　1. *Description* of the Rarities belonging to the
Royal Society preserved in Gresham College. Lond.
1681. fol. pp. 386. pl. 22.

　2. *Rarities* of Gresham College. Lond. 1681.
Wood inserts this, in his Catalogue, as a distinct
work from the former: we have as yet not seen it.

GRONOVIUS, JOHN FREDERICK. — *Ichthyology.*

Cuvier merely says of this writer, that he is the author of several memoirs on fishes, published in various
Transactions.

GRONOVIUS, L. T.— *General Zoology.*

An officer of the municipality of Leyden, who employed an affluent fortune in forming large collections
in different branches of zoology. He was born in 1730,
and died in 1777.

1. *Museum Ichthyologicum.* Leyden, 1754. 1 vol. folio.

2. *Zoophylacium* Gronovianum. 3 parts, forming 1 vol. folio, 1765—1787, with many admirable figures of fishes.

GUALTIERI, NICHOLAS. — *Conchology.*

Professor at Pisa, and subsequently a physician of repute in Florence. His collection of shells must have been valuable.

Index Testarum Conchyliorum quæ adservantur in Museo R. Gualtieri. Florence, 1742. folio. The plates are 110 in number, of which the last four represent *Echini.* The figures, upon the whole, are drawn with great spirit, but are often deficient in those minute details which should characterise the drawings of a zoological artist: hence many of the smaller species cannot be easily recognised. The descriptions are short and unsatisfactory.

GUERIN, F. — *General Zoology.*

A professional draftsman of natural history in Paris, an able entomologist, and a most assiduous naturalist.

1. *Iconographie* du Règne Animal, ou Réprésentations d'après Nature de l'une des Espèces les plus rémarquables, et souvent non encore figurées, de chaque Genre d'Animaux. Royal 8vo. 45 numbers, of 10 plates each. Paris, 1829. A useful and not inelegant work. Most of the figures of the annulose and soft animals are admirable and very correct: but those of the vertebrated class cannot be commended ; they have all the characteristic defects of the French school, and are stiff, disproportionate, and unnatural. M. Guérin has also published, both as editor and part author, the three following works, which we have not seen.

2. *Magazin de Zoologie*, Conchyliologie, et Entomologie ; ou, Description et Figures de Mollusques
vivans et fossiles, et d'Insectes, inédits ou non encore
figurés. 8 vols. 8vo. Paris, 1831—1835. After the
first volume, the title was changed as follows: —
Magazin de Zoologie; Journal destiné à faci-
liter aux Zoologistes de tous les Pays les Moyens de
publier leurs Travaux et les Espèces nouvelles ou peu
connues qu'ils possédent.

3. *Bulletin Zoologique*, Complément du Magazin
de Zoologie ; ou, Annonce et Analise de tous les Ou-
vrages et Mémoires qui se publient sur la Zoologie,
l'Anatomie, et la Physiologie Comparée. Commenced
in 1835, in monthly numbers, 8vo.; and still, we
believe, continued.

GUILDING, THE REV. LANSDOWNE. — *Zoology and
Botany.*

A most accomplished naturalist, and admirable painter,
who resided for many years in St. Vincent's, where he
held the office of colonial chaplain. Unfortunately, he
has left no separate work to record his great talents ;
since most of his papers are in the Linnæan Transactions and the Zoological Journal. His unpublished
MSS. and drawings remain with his widow, now residing in the West Indies. Whatever he touched, he
fully elucidated.

GULDENSTEDT, J. A. — *Zoological Traveller.*

This author is little known in Britain, except by
his name being occasionally attached to species. Cuvier
says, he was born at Riga in 1745 ; was a travelling naturalist in the Russian service from 1768 to 1775 ; and
died at St. Petersburgh in 1781. Besides his *Travels*,
published in German (2 vols. 4to. Petersburgh, 1787
—1791.), he seems to have written several memoirs

in the Petersburgh Transactions, of which we know nothing.

GUNTHER, F. C. — *Ornithology.*

Collection des Figures de Nids et d'Œufs de dif-férens Oiseaux, tirés des Cabinets de Smidel et de celui de l'Auteur. Nuremberg, 1777. folio, pp. 64. pl. col. 101. Complete copies of this book, says Wood, are rare. The last plate represents broken eggs.

GYLLENHAL, L. — *Entomology.*

A learned entomologist of Sweden, whose writings are much quoted.

Insecta Suecica. 4 vols. Lip. 1827. He has also assisted Schœnherr in his work on the Curculionida.

HAGENBACH, J. J. — *Entomology.*

Although this author is known only as having de-scribed a single species of insect, it is nevertheless one of the most extraordinary of the coleopterous order. Cuvier says he was attached to the Royal Museum of Leyden, and died in 1826.

Mormolyce Novum Genus. 1 vol. 8vo. with a plate. Nuremb. 1825.

HAHN, DR. CARL WILHELM. — *Ornithology.*

Voegel aus Asien, Africa, America, und Neuholland, in Abbildungen nach der Natur mit Beschreibungen. (Coloured Plates of the Birds of Asia, Africa, and New Holland.) Two numbers of this work (but little known in England) appeared in 1818; three in 1819; and three in 1820. They contain 54 coloured figures, sufficiently accurate to be easily recognised, and are so far very useful; although the nomenclature is not very accurate.   The descriptions are in Latin and German.

HAMMEL, A. D. — *Entomology.*

1. *Entomological Essays,* No. 1—6. 8vo. Petersb. 1821—27.
2. *Blatta Germanica,* Observations on the. 8vo. Petersb. 1821.

HARDWICKE, MAJOR-GENERAL. — *Zoology.*

A distinguished general officer of artillery in India, where he commanded the whole of that service for many years.   He was passionately attached to natural history, and had native artists constantly in his employ, drawing birds and other animals.   His professional duties, however, did not allow him much time for study.   On his return to England, he placed his immense collection of drawings in the hands of Mr. J. E. Gray, with ample funds for their publication.   By some misunderstanding, however, this object was in a great measure defeated. Meantime the General died, about 1835, at an advanced age, and after leaving, with his accustomed liberality, all his collections and drawings to the British Museum. His executors, unable to adjust the dispute about the copyright, &c. of the work with the editor, threw the whole business into Chancery, where the property will probably remain until it rots.   Neither party will thus derive any benefit; and they are altogether prohibited from selling a single copy of the work in dispute.

1. *Illustrations of Indian Zoology,* consisting of coloured Plates of new. or hitherto unfigured Indian Animals, from the Collection of Major-General Hardwicke.   Selected by J. E. Gray.   12 parts, 10 coloured plates in each.   London: folio.   Many new animals are figured, particularly among the reptiles. The birds and quadrupeds are chiefly taken from the drawings of the native artists.   From the above cause,

probably, no descriptions have been published. That general Hardwicke was fully competent to have undertaken the publication of his own materials, will be apparent, by the following papers he contributed to the Linnæan Transactions ; but, on his return to England, he was far advanced in years, and naturally felt adverse to the trouble of superintending such a work.

2. *Description of Mus giganteus.* Linn. Trans. vol. ix. p. 115.

3. *Description of the Wild Dog of Sumatra,* and a new Species of Viverra. Ib. xiii. p. 23.

4. *Description of Two new Species of Antelope ;* with 3 plates. Ib. xiv. p. 518.

5. *Description of a new Indian Bat ;* with a plate. Ib. xiv. p. 525.

6. *Description of a new Genus of Mammalia* (*Ailurus* F. Cuv.). Ib. xv. p. 161.

7. *Account of Ursus Indicus,* the Indian Badger. Ib. ix. p. 379.

HARLAN, DR. RICHARD.—*Erpetology.*

distinguished American erpetologist and physician, Professor of Comparative Anatomy in Philadelphia. Besides many valuable contributions to the learned societies of his own country, he has published —

Fauna Boreali Americana. Philadelphia, 1825. 1 vol. 8vo. The first, containing the quadrupeds, of a projected general work on North American zoology.

HARRIS, MOSES.—*Entomology.*

The best painter and engraver of insects at the period in which he lived, besides being a most accurate describer. Although not an educated man, he was an accurate and original observer ; for he was the first, we believe, who called the attention of entomologists to the neuration of the wings, and even arranged his genera conformable thereto. Besides the plates to Drury's insects, and the engraving of those to Smith and Abbot's, he is the author of the following works : —

1. *The Aurelian,* or Natural History of English Butterflies and Moths. 1 vol. folio, with admirably coloured plates. Original copies are now very rare and valuable. A new edition has been recently published ; but the plates are worn down, and the colouring is too gaudy.

2. *An Exposition* of English Insects (chiefly confined to the Neuroptera, Hymenoptera, and Diptera) ; with 51 coloured Plates, and near 500 Figures. 1 vol. 4to. London, 1782.

HARTMANN. — *Zoological Artist.*

A zoological artist and engraver, residing at St. Gall; author, says Cuvier, of a system of the terrestrial and fluviatile *Testacea* of Switzerland.

HASSELQUIST, FREDERICK. — *Zoological Traveller.*

Frederick Hasselquist was one of the numerous pupils of Linnæus, in whom the instructions of that great master had excited such an enthusiasm for natural history, that they gladly availed themselves of every opportunity of visiting foreign lands in search of new objects. He was likewise one of the many so circumstanced, who were cut off in the midst of their usefulness, and whose premature fate led Linnæus to exclaim, " Surely no science has had so many martyrs as natural history."

He was born on the 14th of January, 1722, at Toernvalle, in Ostrogothia, where his father was a vicar. His family being very poor, Hasselquist was left, at his death, in very destitute circumstances ; but, by the assistance of some generous friends, he was enabled to attend the university of Upsal (1741), where he supported himself chiefly by private teaching. His love for natural history was fostered by Linnæus, who was not slow to discover that nature had endowed him with superior powers. Happening to be present at a lecture, in which the illustrious professor took occasion to express his regret that so little was known about the vegetable productions of Palestine, he conceived a strong desire to visit that country, and supply the desideratum. Many difficulties stood in the way ; but these were at last overcome ; and, in the spring of 1749, he repaired to Stockholm, where he obtained a passage to Smyrna in an East Indiaman. He reached that place in the end of November in the same year : he then visited Egypt, residing some time at Cairo, surveying the pyramids, &c. ; returned to Palestine by way of Damietta and Jaffa ; and, joining a caravan of pilgrims, at last arrived at Jerusalem. Having in some measure satisfied his curiosity regarding a spot where there is so much to engage the attention and affect the heart, he visited the banks of the Jordan, Jericho, Bethlehem, Rama, Nazareth, Tiberias, Tyre, and Sidon, &c. During the whole of this journey, it is scarcely necessary to state, he took every opportunity of collecting plants ; and not only these, but objects in every department of natural history, as well as manuscripts, antiquities, &c. He then returned to Smyrna ; touching at Rhodes, Cyprus, and Scio, in his passage, with the intention of embarking from that port for Europe. His various journeys and unceasing exertions had, however, proved too much for a frame naturally weak ; and, while waiting at Smyrna to recruit his strength, he was carried off by pulmonary consumption, on the 9th of February, 1752, in the thirtieth year of his age. His pecuniary means, which were never ample, had been exhausted some time previously, and his collections remained in the hands of his creditors. On this becoming known in Sweden, the property was ordered to be redeemed by queen Louisa Ulrica, to whom it was accordingly transmitted. Much of it came into the hands of Linnæus ; and he lost no time in publishing the manuscripts of his deceased friend, under the title of *Iter Palestinum*. The first part of this work contains the traveller's journal ; the

second, his observations on natural history, arts, manufactures, diseases, &c. A Flora of Palestine, drawn up
from his papers and specimens, was afterwards laid before the public. — J. D.

HAYES. — *A Zoological Artist.*

Figures of rare and curious Birds, accurately
drawn from living Specimens, and faithfully coloured;
with a descriptive Account of the Habitation, Character, and Qualities of each Subject. London, 1822.
4to. Published in monthly numbers, each containing 4 plates. The prospectus states the whole will
not exceed 25 numbers.

HERBST, J. F. W. — *Entomology.*

A zealous entomologist, and industrious compiler and
writer. Born at Berlin, in 1743.

1. *Natursystem* aller bekannten Insecten, i. e.
The Natural History of all the known Insects, Native and Exotic (in German). Berlin, 1782—1806.
21 vols. of text, 8vo., and 21 of plates, 4to. This
extensive work is divided into two parts. The first
comprehends 10 volumes, each accompanied by one
of plates, and contains the Coleopterous Insects (Der
Käfer). The second is of 11 volumes of letterpress and plates, and includes the *Lepidoptera* (Der
Schmetterlinge). Very few of the figures are original.
It is the joint work of Herbst and C. G. Jablonsky.

2. *Versuch* einer Naturgeschichte der Kraben, &c.,
i. e. The Natural History of Crustacea. Berlin,
1790—1803. 3 vols. 4to. pl. col. 62. A useful compilation, containing several figures of new species.

3. *Natursystem* de ungeflügelten Insecten, &c.
The Natural History of the Genera Solpuga, Taran-
tula, and Phalangium. Berlin, 1797. 1 vol. 40 pl. col.

4. *Genus Scorpio,* Natural History of the. Berlin,
1798.

HEERKENS, G. N.— *Ornithology.*

Aves Frisicæ.   Rotterdam, 1787. pp. 298. pl. 1.

HERMANN, JOHN.— *General Zoology.*

A learned and laborious naturalist, who was long professor at Strasburg. He was born in 1738, and died in 1800.

   1. *Tabula Affinitatum* Animalium.   Strasb. 1783. 1 vol. 4to.

   2. *Observationes* Zoologicæ Posthumæ.   Strasb. and Paris, 1804. 1 vol. 4to.

HERMANN, J. F.— *General Zoology.*

Son of the professor of that name.   Born in 1768; died in 1793.

   Mém. Aptérologigue.   Strasb. 1804. 1 vol. folio. (*Cuvier.*)

HERNANDEZ, FRANCESCO.

A physician, who was sent out by Philip II. to Mexico, with a liberal salary, and every facility given him to investigate the productions of that country. It is to be lamented that such munificence should have been attended with little or no advantage to science ; for Hernandez had neither the talents nor knowledge to fill such a situation.   His book is crude, and without the least merit.   He describes, indeed, a vast number of objects ; yet in such vague terms, that they cannot be identified.

   Nova Plantarum, Animalium, et Mineralium Mexicanorum Historia.   Roma, 1651. folio.

HOFFMANSEG, COUNT. — *Zoology and Botany.*

An ardent and accomplished zoologist, who devoted a large part of his fortune (which he much impaired) to his favourite pursuits.   He sent collectors, both zoologists and botanists, to Brazil and Portugal; but confined his writings to essays and papers in various Transactions.

HOME, SIR EVERARD. — *Surgery and Comparative Anatomy.*

A celebrated comparative anatomist, but too prone to theorise upon other subjects he did not understand. It has been confidently said, — with what truth we know not, — that he made use of the Hunterian MSS. without acknowledgment.   Of zoology, in general, he knew but very little.

HOPE, THE REV. FRED. WILL. — *Entomology.*

An eminent coleopterist of this country, and possessor of a large collection of insects.   Besides being the author of near 30 papers, in Transactions and periodicals, he has published separately —
    1. *The Coleopterist's Manual;* in three parts, 8vo. London, 1837—9 ; a highly interesting work.
    2. *Hemiptera,* Catalogue of, in the author's Collection.   London, 1837.

HOPPE, D. H. — *Entomology.*

An apothecary of Ratisbon, who seems to have paid great attention to the coleopterous insects of his native province.
    Enumeratio Insectorum Elytratorum indigenorum. Erlangæ, 1795. 1 vol. 8vo. with a few coloured plates, the figures of which are very excellent.

HORSFIELD, DR. THOMAS. — *General Zoology.*

A celebrated zoologist and botanist, who resided many years in Java as a physician, where he made large col-

lections both of its plants and animals. On the taking
of Java by the British, under sir Stamford Raffles, he
was induced to enter the service of the Honourable East
India Company ; and is now placed in charge of their
rich and valuable museum.

1. *Zoological Researches* in Java and the neighbouring Isles. London, 1825. 1 vol. 4to. with numerous beautiful plates.

2. *Descriptive Catalogue* of the Lepidopterous Insects in the Museum of the Honourable East India
Company ; illustrated by coloured Figures of new
Species, and of the Metamorphosis of Indian Lepidoptera, with Observations on a general Arrangement of this Order of Insects. London, 1829. Royal
4to., with some of the most elaborate and beautifully
coloured plates that have ever been published.   Two
parts only have yet appeared.   The Introduction is
full of important and interesting observations, highly
conducive to the advancement of science. Dr. Horsfield has also contributed largely to the scientific
journals, the Linnæan Transactions, Zoological Journal, &c.

HOUTTUYN, MARTIN.—*Ornithology.*

Besides several papers printed in the Haarlem Transactions, he continued and completed Noseman's *Birds
of the Netherlands.*

HUBER, FRANCIS.— *Entomology.*

On perusing the lives of naturalists, we not unfre-
quently meet with instances of indefatigable perseverance under the most discouraging circumstances, and
of important discoveries being made, when there could
be very slight prospect of effecting them on the part of
the inquirer.   But of all the investigators of the minute
operations of living beings, the case of Francis Huber

seems the most hopeless; for he laboured under *total blindness,* — a defect which might be thought sufficient to paralyse every exertion, and render every attempt at research nothing less than a ludicrous absurdity. Yet such was his sagacity and skill in planning and executing experiments, that we are indebted to him, more than any other writer, for an explanation of many of the mysteries of the hive-bee, — the subject to which he almost exclusively applied himself.

He was born at Geneva, on the 2d of July, 1750. His father, John Huber, was a man of a very lively disposition, being esteemed the greatest wit of his neighbourhood: as such, his society was sought after by Voltaire, who did not think his sayings and verses unworthy of being repeated to the élite of Ferney. He was likewise much attached to natural history, and published a treatise on the flight of birds of prey. His son Francis inherited his father's disposition and predilections. In his younger days, he enjoyed the instructions of Saussure, and became such a diligent student, that his health was affected by too earnest application. His eyesight was also injured; chiefly from the practice of sitting up greater part of the night, reading romances, often by the light of the moon, after his taper had failed him. At this time he was about the age of fifteen. His father took him to Paris, to consult Venzel, the most celebrated oculist of the day; but his case was declared to be a hopeless one. Before his vision was completely lost, he had secured the affections of Maria Aimée Lullin, daughter of a Swiss magistrate; and, although their union was strongly opposed by the lady's father and relations, she no sooner came of age, than she united her fate with that of Huber, although he was now completely blind. It was through her devoted affection that he was enabled to accomplish so much: she was his reader, secretary, observer; literally becoming 'eyes to the blind.' His servants, also, particularly Francis Burnens, afforded him the utmost assistance; and, latterly, his son Pierre, who distin-

guished himself in the same kind of inquiries which his
father prosecuted so successfully.

Huber's "Nouvelles Observations sur les Abeilles,"
in 1 vol. 8vo., appeared in 1792. It is written in the
form of letters to his friend Bonnet, the well-known
author of the "Contemplation de la Nature." It contains many important discoveries; and the confirmation
of many facts, before his time most imperfectly known.
The most important relate to the impregnation of the
queen bee—the conversion⁻of a worker larva into a
queen —origin of wax—nature of propolis—ventilating
hives, &c. The work was so well received, and considered so extraordinary in the peculiar circumstances
of the author, that he received the honour of being admitted into most of the learned societies of Europe, and
was celebrated in verse by the poet Delille.

During the latter years of his life, after the decease
of his devoted and generous-minded wife, he resided
with his married daughter, madame de Moulin, at Lau-
sanne ; and it was there that he died, on the 22d of December, 1831, having attained the advanced age of
eighty-one. — J. D.

Nouvelles Observations sur les Abeilles. Second ed.
Paris and Geneva, 1814. 2 vols., the second of which
is by his son Pierre.

HUBER, PIERRE. — *Entomology.*

Recherches sur les Mœurs des Fourmis indigènes.
Paris and Geneva, 1810. 1 vol. 8vo. with figures.

HÜBNER, J.— *Entomology.*

A painter of natural history at Augsbourg. In his
great work on the European *Lepidoptera*, he has succeeded in imitating the delicate tints of these insects
with a truth and accuracy far surpassing any of his
predecessors. The number of species are likewise more
numerous. The text is in German. Hübner seems to

have meditated a similar publication on the exotic *Lepidoptera :* several plates are said to have been published, but these we have not seen.

Der gamlung Europaïschen Schmetterlinge. Augsb. 1796. 4to.

HUET.—*Zoological Painter.*

One of the artists attached to the Royal Museum at Paris.

Collection de Mammifères du Muséum d'Histoire Naturelle, dessinés d'après Nature. Paris, 1808. 4to. pp. 60. col. pl. 52.

HUMBOLDT, ALEXANDER BARON DE.

This illustrious traveller was born at Berlin, in 1769. Scarcely any branch of science has been left unenriched by his discoveries. We can only regret that zoology engaged so little of his attention during his memorable travels in Tropical America.

Observations de Zoologie et d'Anatomie Comparée. Paris, 1811. 1 vol. 4to. col. pl.

HUMPHREY, GEORGE.— *Conchology.*

Celebrated for being the first reformer of conchological arrangement, after the system published by Linnæus. With a mind so evidently capable of working out the system he only promulgated, it is to be regretted that professional pursuits, and the baneful spirit of procrastination, acted as a bar to his leaving behind him any descriptive work. As an author, therefore, he is little known ; since the remarkable pamphlet which developed his views, was published anonymously. Mr. Humphrey, for many years, was the chief commercial naturalist in this country; and from his father, who was in the same profession, he inherited immense collections

both in conchology and mineralogy. With a disposition
the most amiable, and manners the most gentlemanly,
yet unassuming, his company was sought for by all the
great collectors and naturalists of his time, such as the
duchess of Portland, the earl of Tankerville, Dr. Fordyce, Mr. Jennings, &c. By these he was esteemed not
merely for his scientific acquirements, but as an humble
friend, with whom they could hold communion " sweet
and large," on their favourite topics. Yet such was his
indifference to payment, that although his remuneration
was always liberal, the worthy man was always straightened for money, if not absolutely poor. A large family,
and a series of domestic misfortunes, no doubt, contributed to keep him under this pressure up to the period
of his death, which happened about 1830. His eldest
son, Adolirius, was one of the early colonists in Van
Diemen's Land; to which place he went, as government
mineralogist, about the year 1803 or 1804. This situation he soon resigned ; and then, entering into agricultural pursuits, finally became one of the most wealthy,
and certainly one of the most respectable, landholders in
that island. His other children, excepting two, being
married, he retired with these to a small house at Chelsea, where he closed a long and blameless life at a very
advanced age. He was, indeed, " an Israelite in whom
there was no guile." Although, from the kindness of
his nature, and the belief that all the world was as honest
as himself, he was perpetually wronged and defrauded,
yet I never heard him say an unkind, much less an
angry word, against any human being. His integrity
was such, that countless bags of unique shells (which
he prized much more than gold) might have been committed to his charge, unsecured by a receipt. The reader
may think, perhaps with reason, that I have said too much
of this good man ; but I *knew* his worth, and cherish
his memory ; for he was my first preceptor and encou-
rager in the study of nature. The quick eye of a parent
foresaw, that if this early passion for natural history
was not suppressed, it would absorb all other pursuits ;

even to the exclusion of any wish for a professional life:
my juvenile ardour was therefore repressed ; and I was
even afraid of exhibiting it at home. When, however, I
could steal an hour to visit, or had permission to spend a
day with Mr. Humphrey, it was the greatest happiness
of my life. All my shells, whose names I knew not, were
thrust into my pockets, and emptied before the good old
man, who, with quiet complacency, would sit down and
write me little tickets for each, in one of the most beautiful and legible hands I ever saw. A few trifling additions were generally made ; and thus, without the least
idea of counteracting the wishes of my parents, that
flame was quietly, and almost imperceptibly, fanned,
which ultimately led me to make all worldly advancement subordinate to the study of nature.   Mr. Humphrey sold the whole of his collections to Mr. G. B.
Sowerby, who subsequently disposed of the greater part
by auction.

 1. *Museum Calonnianum.*   Specification of the
various Articles which compose the magnificent Museum of Natural History collected by M. De Calonne,
in France.   Published anonymously.   London, May
1. 1797.

 2. *On the Animal of Bulla lignaria.*   Printed in
vol. i. of the Linnæan Transactions.   If we do not
mistake, this was the first exposition of the fraud, as
Cuvier has called it (but more likely the error), of
Gioeni, who described the testaceous stomach of this
animal as a new and distinct animal, and it is figured
as such in the *Ency. Méthodique.*

HUNT.— *Ornithology.*

A zoological artist of Norwich, whose work is now
seldom met with.

 British Ornithology, containing Portraits of all the
British Birds ; including those of Foreign Origin
which have become domesticated.   Norwich, 1815
—22. 8vo. 13 parts already published, with 12 co-

loured plates in each.   To those who are desirous of
possessing coloured plates of our native birds at a
moderate price, we recommend this work. The figures
are easy and characteristic, slightly but faithfully
coloured, and the text adapted both for popular and
scientific use.

HUNTER, DR. JOHN.— *Comparative Anatomy.*

We are not aware of any distinct work on zoology
written by this celebrated anatomist; but to him the
College of Surgeons is indebted for its splendid museum.   He was born in 1728, and died in 1793.

ILLIGER, J. CH. WILLIAM. — *General Zoology.*

One of the first zoologists of his age; he was Professor at Berlin, but unfortunately died at an early age.
Whether we consider the value or the extent of Illiger's
labours, we feel surprised that, at an age comparatively
early, he should have done so much.

1. *Verzeichniss* der Kæfer Preussens, i. e. Descriptive Catalogue of the Coleopterous Insects of Prussia,
commenced by T. Kugelann, and finished by Illiger.
Halle, 1798. 1 vol. 8vo.

2. *Systematisches Verzeichniss* von den Schmetter-
lingen der Wiener gegand, i. e. Systematic Catalogue
of the Lepidopterous Insects of Austria.   Bruns.
1801. 2 vols. 8vo.

3. *Magazin für Insectenkunde.*   Bruns. 1801—
1807. 7 vols. 8vo.

4. *Prodromus* Systematis Mammalium et Avium.
Berlin, 1811. 1 vol. 8vo.   From this admirable and
philosophic work M. Vieillot has drawn largely. Besides the above, Illiger continued the edition of Hell-
wigg, of the *Fauna Etrusca*, the second volume of
which was published at Helmstad, 1807. 8vo.

JACOB, N. H. — *Mammalogy.*

Histoire Naturelle des Singes, où chaque Espèce
est representée accompagnée d'un Texte Italien, Française, et Allemande. Milan, 1812. folio. pp. 91.
pl. 70.

JACQUIN, NICHOLAS AND JOSEPH. — *General Natural
History.*

Two celebrated botanists and travellers, father and
son, the authors respectively of the following zoological
papers : —
    1. *Miscellanea Austriaca.*   Vienna, 1778—1781.
2 vols. 4to.
    2. *Ornithological Essays* (in German). Vienna,
1784. 4to.

JARDINE, SIR WILLIAM, BART., AND P. J. SELBY. —
Ornithology.

Two of our best known ornithologists, who have described and figured many new and interesting birds in
their joint work, entitled —
    Illustrations of Ornithology. Edinburgh, 1829, &c.
3 vols. royal 4to. Sir William Jardine is also the
editor of a small popular periodical work, entitled,
The Naturalist's Library, with coloured plates, of
which many volumes have been published. Both
these gentlemen are editors of the *Magazine of
Zoology and Botany,* 2 vols. ; now continued under
the title of *Annals of Natural History.* It is the
best periodical of its kind now published ; and appears in monthly numbers, of which their are already 30.

JOHNSON, DR. GEORGE. — *Malacology.*

An eminent physician, settled at Edinburgh, and one of the most learned and accomplished malacologists of this country.

History of British Zoophytes. 1 vol. 8vo. London, 1839. With many excellent figures. It is to be hoped that the numerous papers of the author, now scattered in the scientific periodicals, will be collected, ere long into a separate volume.

JOHNSTON, JOHN. — *General Zoology.*

A laborious naturalist of the seventeenth century. Born at Sambter, in Poland, in 1603; died in 1675. His writings are very numerous; but we shall only enumerate the best of those relative to zoology.

1. *Historia Naturalis* de Quadrupedibus, Avibus, Piscibus, Insectis, et Serpentibus, Libri vi., cum æneis Figuris. Johannes Johnstonis, M.D., concinnavit. Franc. 1650. 2 vols. folio.

2. *Piscibus et Cetis,* &c., Historia Naturalis de. Francf. 1649. folio.

3. *Insectibus,* Historia Naturalis de. Francf. 1653. folio.

4. *Animalium,* Historia Naturalis, cum Figuris. Amst. 1657. 2 vols. folio.

JURINE, LOUIS. — *Entomology.*

Professor of Anatomy and Surgery at Geneva: an acute and learned entomologist, whose observations have thrown great light upon the hymenopterous insects.

Nouvelle Méthode de classer les Hymenoptères et les Diptères. Geneva, 1807. 4to. coloured plates. Of this admirable work, the first volume, containing the *Hymenoptera,* was alone published. The plates are of great beauty, and are executed with uncommon fidelity. The idea, however, of classing these insects by the nerves of their wings, was not new ; for it had long before been done by our countryman, Harris.

KÆMPFER, E. — *Zoological Traveller.*

A learned physician and naturalist, who travelled in Persia and various parts of India : he wrote an account of Japan, in 2 folio volumes, which is still consulted. Born in 1651 ; died in 1713.

Amœnitatum Exoticarum Fas. 5. Lemgo, 1712. 4to. (*Cuv.*)

KIRBY, THE REV. WILLIAM. — *Entomology.*

One of the most celebrated entomologists of the present day; whose popular *Introduction,* in which he was assisted by Mr. Spence, has procured him considerable fame. His work on English bees is considered a model for similar investigations. Mr. Kirby was elected the first President of the Entomological Society of London ; to which he subsequently presented his fine collection of insects.

1. *Monographia* Apum Angliæ. 2 vols. 8vo. Ipswich, 1802.

2. *Introduction to Entomology,* by Messrs. Kirby and Spence. 4 vols. 8vo. London, 1828, &c. With numerous anatomical engravings. Mr. Kirby's other valuable writings are scattered in various Transactions and periodicals.

KLEEMANN, C. F. C. — *Entomological Painter.*

An artist of Nuremberg. Born in 1735; died in 1789. He was a relation of Rœsel's, and published the 5th, or supplementary, volume to that author's work on insects.

Beyträge zur Natur oder Insecten-geschichte. Nuremberg, 1761. 4to.

Q

KLEIN, J. TH. — *General Zoology.*

Born in 1683. He was secretary to the senate of
Dantzig. Although an assiduous and indefatigable
writer on every branch of natural history, yet he had not
much talent or judgment. He was one of the most determined, and yet one of the weakest, opponents of the
great Linnæus ; and his nomenclature and characters are
consequently obsolete.

1. *Quadrupedum* Dispositio brevisque Historia
Naturalis. Lips. 1751. pp. 127. pl. 5.

2. *Historiæ* Avium Prodromus. 1750.

3. *Stemmata* Avium, Latine et Germanice. Lips.
1759. 4to. pp. 48. pl. 40.

4. *Summa* dubiorum circa Classes Quadrupedum et
Amphibiorum Linnæi. 1743.

5. *Naturalis* Dispositio Echinodermatum. 1734.

6. *Descriptiones* Tubulorum Marinorum. 1737.

7. *Mantissa* Ichthyologica. 1746.

8. *Historiæ Nat.* Piscium promovendæ Missus 5.
1740—1749.

9. *Methodus* Ostracologica. 1753.

10. *Tentamen* Herpetologiæ. 1755.

KLIENER. — *Conchology.*

A young and very promising naturalist, curator of the
prince of Massina's museum in Paris.

Species Général et Iconographique des Coquilles
vivantes, comprenant le Musée Massena, la Collection
Lamarck, celle du Muséum d'HistoireNaturelle, et des
découvertes récentes des Voyageurs. Paris, in 4to.
and 8vo.; published in quarterly numbers, of six
coloured plates, with descriptions, and of which 42 have
already appeared. Although few can live to see the
termination of this *General Conchology*, it well deserves patronage, the figures being very good, and the
price moderate.

Klug, F. — *Entomology.*

Doctor of Medicine, and Professor in the Museum of Berlin. An entomologist of considerable authority, who has added much to our knowledge of many families in the order *Hymenoptera.*

Monographia Siricum Germaniæ, atque Generum illis adnumeratorum.   Berolini, 1803. 1 vol. 4to. with 8 col. plates.   Many other essays by Dr. Klug will be found in scientific journals.

Knoch, A. G. — *Entomology.*

Neue Beytraege zur Insectenkunde, &c., i. e. New Materials for a Knowledge of Insects. Leips. 1801. 1 vol. 8vo. fig.

Knorr, G. W. — *Zoological Painter.*

A painter and engraver of natural history, of some merit.   Born in 1705 ; died in 1761.

Les Délices des Yeux et de l'Esprit, ou Collection générale des différentes Espèces de Coquillages que la Mer renferme.   Nuremb. 1760—73. 4to. 3 vols. in 6 parts.   The figures, in general, are accurate ; the text is poor.

Kuhl, Henry. — *General Zoology.*

A young but very accomplished zoologist, the companion of Van Hasselt in their unfortunate expedition to Java.   He published several valuable papers, in German, previous to his leaving Europe. (See Van Hasselt.)

Laet, Jean de. — *General Zoology.*

An esteemed geographer of the 17th century.

Novus Orbis, seu Descriptionis Indiæ Occidentalis Libri XVIII.   Leyden, 1633. folio.

LACEPÈDE, BERNARD-GERMAIN COUNT DE.—*Ichthyology.*

One of the most celebrated ichthyologists since the
days of Artedi.

1. *Histoire Naturelle,* générale et particulière, des
Quadrupèdes Ovipares et des Serpens. Paris, 1788
—89. 2 vols. 4to.

2. *Poissons,* Histoire Naturelle des. Paris, 1798
—1803. 15 vols. 4to.

3. *Cétaces,* Histoire Naturelle des. Paris, 1804.
1 vol. 4to.

LAICHARTING, J. N.—*Entomology.*

Professor at Inspruck. Died in 1754.
Verzeichniss der Tyroler Insecten. Zurich, 1781
—1784. 2 vols. 8vo.

LAMARCK, J. B. CHEVALIER DE.—*Malacology.*

The full name of this illustrious man, distinguished for
his intimate acquaintance with the invertebrate animals,
and his admirable skill in the perception of natural affinities, was Jean Baptiste Pierre Antoine de Monet, but
he was generally styled the Chevalier de Lamarck. He
was descended from a family of some rank; and born at
Bezantin, a small village in Picardy, on the 1st of
August, 1744. The family being very numerous, and
his father's circumstances somewhat reduced, the subject
of the present notice was destined to obtain a livelihood
in the church; and was placed under the Jesuits at
Amiens, that his education might be conducted with
that view. He conceived, however, an utter dislike to
a college life; and on the death of his father, which
took place in 1760, joined the French army then in
Germany, as a volunteer. Having behaved with the
utmost bravery in an action which took place at Fis-

singhausen, the very day after his arrival, he was promoted to a lieutenancy; but the bright prospects which
this success afforded, were speedily overcast by an accident which compelled him to abandon the army for
ever. He repaired to Paris, where he obtained a scanty
subsistence, for a time, by acting as a clerk. At the
same time, however, he eagerly resumed the physical
studies which he had done little more than commence
at college, and turned his views to the medical profession.

Botany and meteorology constituted the subjects of
his earliest studies; and his first work of importance was
a " French Flora, or a brief Description of all the Plants
which grow naturally in France,"—the plants arranged
upon a binary or dichotomous system. This work was
the means of introducing its author to Buffon, then in
the height of his popularity, through whom he obtained
a situation in the botanical department of the Academy
of Sciences. The Count afterwards conferred upon him
a still more signal advantage, by appointing him tutor
to his son, when he was about to set out to make the
tour of Europe; superadding a commission as botanist
to the king, which gave him a kind of official character,
and procured him a friendly reception from all the botanists he encountered in his route. After his return,
he made extensive contributions on botany to the *Encyclopédie Méthodique*. At last he was appointed to a
permanent situation, by the influence of M. De la Bil-
lardière, a relation, by which the duty was assigned
him of keeping the herbaria in the king's cabinet.
When the Museum of Natural History was established,
in 1793, he obtained the appointment of Professor; his
duty being to lecture on the two classes of the animal
kingdom named *Insecta* and *Vermes* by Linnæus. On
the study of this department, which had not previously
attracted a large share of his attention, he entered with
his characteristic zeal; and the result was, his invaluable
work entitled *Histoire Naturelle des Animaux sans Vertèbres*. The first five volumes of this work were written

entirely by himself, with the assistance of Latreille in
the portion relating to insects. A part of the sixth was
composed by M. Valenciennes; and the whole of this
volume, as well as the seventh, was prepared for the press
by his daughter, — his loss of sight preventing him from
undertaking the task himself. His " General Division
of Animals" first appeared in 1812, in a small volume
purporting to be an extract from his course of lectures.
He afterwards divided animals into two sub-ramose
series, adding a new class, viz. the *Ascidiens*. This distribution is admitted by Mr. MacLeay to be the first
approach to a perception of that order of affinities which
pervades the animal kingdom; and nearly coincides
with the tabular view which he himself laid before the
public. Lamarck, amid the multitude of subjects which
engaged his mind, paid much attention to the history of
fossil shells, for the study of which the vicinity of
Paris presents an excellent field. The result of his
investigations appeared in the earlier volumes of the
" Annals of the Museum."

His eyes had long been weak, and during many of
the latter years of his life he was totally blind. This
calamity was superadded to many reverses of fortune,
and his declining years were spent in comparative poverty. He had been married four times, and was the
father of seven children ; some of whom, and particularly his eldest daughter, tended him during his infirmity with the most unwearied filial affection. He
died on the 18th of December, 1829, in the 85th year
of his age.

The limits to which these biographical notices must
necessarily be confined, prevent us explaining, at such a
length as is requisite to make them at all intelligible,
the various theories Lamarck formed on many of the
great phenomena of nature, and his method of ac-
counting for the varied forms in which animal life now
appears. His speculations on these subjects may be
briefly characterised, not merely as fanciful, but absolutely
absurd ; leading, in some instances, if legitimately fol-

lowed out to their conclusions, to consequences of a very
pernicious tendency. They must be admitted, however,
to display no small degree of ingenuity, and a fund of
varied knowledge indicative of a capacity of a superior
order. — J. D. W. S.

   1. *Systême des Animaux* sans Vertèbres. Paris,
1801. 1 vol. 8vo.   This prelude to the enlarged work
is now become very rare.

   2. *Extrait du Cours de Zoologie* sur les Animaux
sans Vertèbres. Paris, 1812. Pamphlet in 8vo.

   3. *Histoire Naturelle des Animaux sans Vertèbres.* Paris, 1815. 7 vols. 8vo.   A new edition of
this valuable work, enriched with a great addition of
species, and valuable notes by Deshayes and Milne
Edwards, is now in course of publication at Paris.

LAMOUROUX, J. V. F. — *Coralline Animals, &c.*

An able naturalist, and one of the Professors at Caen.

   1. *Histoire des Polypiers,* 1 vol. 8vo. plates.

   2. *Exposition* Méthodique de l'Ordre des Polypiers;
with the plates of Ellis and Solander, and some new
ones. Paris, 1821. 1 vol. 4to.

   3. *Dictionnaire* des Zoophytes ; forming part of
the Ency. Méth. Paris, 1824. 1 vol. 4to.

LANGSDORFF, G. H. VON. — *Zoology and Botany.*

An enthusiastic traveller and zealous collector, who
accompanied the Russian circumnavigator, admiral
Krusenstern, round the world, and was subsequently
appointed Russian consul-general to the Brazils.   M.
Langsdorff has not, we believe, written any distinct
zoological work, although he possessed a very fine collection of Brazilian insects.

   Voyages and Travels in various Parts of the World
during 1803—1807. London, 1813—14. 2 vols.
4to., in which are many zoological observations. The

author fell a victim to fever, caught when exploring
the sources of the great river Tocantine, a branch of
the Amazon.

LA PEYROUSE, PHILLIPE PICOT, BARON DE.—
Conchology.

Professor of Natural History at Toulouse.
Description de plusieurs Espèces d'Orthoceratites
et d'Ostracites.   Nuremb. 1781. folio.

LASPEYRES, J. H.—*Entomology.*

One of the municipal officers of Berlin.   An acute
and meritorious entomologist, who has paid particular
attention to the Linnæan *Sphingidæ,* one division of
which he has investigated with consummate ability.

Sesia Europeæ Iconibus et Descriptionibus illustratæ.   Berolini, 1801. 4to. 1 coloured plate.   The
figures are remarkably accurate, and the species
clearly defined.   Entomologists expect to be gratified,
shortly, by another work on this interesting family,
from the pen of this writer.

LATHAM, DR. JOHN.—*Ornithology.*

The works of Latham will be long quoted, because,
although exhibiting more of unwearied zeal, and extensive research, than of critical acumen or comprehensive
judgement, they have become interwoven with the science
he cultivated, and are cited by almost every writer.   Although a strict disciple of the Linnæan school, ánd
hence strongly prejudiced against the growing innova-
tions upon his master's nomenclature, which our Continental neighbours were even then making, he was so
far unprejudiced as to characterise several new genera
(a bold step in those days), and to separate the land
from the aquatic birds.   This was certainly an advance ;
although a small one.   For the rest, we are obliged to
say that the vastness of his plan, which aimed at no less

than the description of all known birds, was too great
for his talents. His memory was not good; hence he
has frequently described the same species by different
names; and he placed too much faith in drawings, which
led to the same error. Dr. Latham happened to live at
that particular period when the museums of Europe
began to be crowded with new birds, quite unknown to
Linnæus, without any one naturalist to describe them.
This he undertook to do; and, in reference to the then
state of ornithological science in Britain, he did his
task very creditably. But he wrote only for his cotemporaries,—not for posterity, or even for that generation,
which he was destined, by a long life, to see. Hence it
is, that, having now so many more accurate writers, Dr.
Latham will only be quoted to explain doubts, or for
the numberless errors in his volumes. His reading was
most extensive; and the research he bestowed to eluci-
date his subjects is far greater than what is now taken
by certain writers, with whom every species they have
not seen before, is characterised as " new to science."
In private life, Dr. Latham was a most amiable man;
and he lived, much esteemed and respected, to a very
advanced age.

  1. *A General Synopsis of Birds.* 8 vols. small 4to.
with coloured plates. London, 1782.

  2. *Index Ornithologicus.* 2 vols. 4to. London,
1790.

  3. *A General History of Birds;* being an enlarged
edition of the General Synopsis, with a few additional
plates, hardly any new divisions, and scarcely any cor-
rections. 11 vols. 4to.

LATREILLE, PIERRE-ANDRÉ.— *Entomology.*

Pierre-André Latreille, whose name marks the
third great era of Entomological Science, was born at
Brives, in the department of Corréze, on the 29th
of November, 1762. His parents were of an honourable family; but, owing, we presume, to their early

death, the prospects of their son were but of a very
indifferent description.   He was fortunate, however, in
meeting with friends and protectors, who took charge of
his boyhood and provided for his education.   The principal of these were a M. Laroche and M. Malepeyre; the
latter of whom was the first to inspire him with a taste
for natural history, by supplying him with books on that
subject.   Another of his early patrons was the baron
d'Espagnac, governor of the Invalides, at whose request
Latreille came to Paris in 1778.   Here he was placed in
the college of cardinal Lemoine, and prosecuted various
branches of education with much success ; his love for
natural history continuing all the while to acquire additional intensity.   It was partly for this reason that he
was honoured with the attention and esteem of the celebrated mineralogist Haüy.   But, during the whole of
this period of his life, he was most indebted to the family of M. D'Espagnac, who seem to have regarded him
as an adopted son ; and even after the decease of that
gentleman, his sister, the baroness de Puymarets, and
others of the family, still continued their acts of kindness to him.   In his twenty-fourth year, he spent a considerable time in the country, and the whole of it was
devoted to the study of insects, to which he was by this
time strongly attached.   We believe that the first fruits
of his researches on this subject—one that occupied
such a large portion of his after life—was a memoir
on the Mutillas (a genus of hymenopterous insects) of
France ; this essay led to his being elected, in 1791, a
corresponding member of the Nat. Hist. Society of Paris,
and shortly after, of the Linnæan Society of London.
An intimacy was likewise established between him and
Olivier, Bosc, Lamarck, and other cultivators of natural
science in the metropolis.   Some of his earliest entomological writings consisted of contributions to the
Encyclopédie Méthodique.
    During the paroxysm of the revolution, M. Latreille
—the more especially as he was a member of the eccle-
siastical body—had his full share in the vicissitudes inci-

dent to that eventful period. He was, in fact, imprisoned; and would, in all probability, have met the fate to which so many other victims, equally innocent, were subjected, but for a singular occurrence, which has often been referred to, and which he himself has commemorated in his *Gen. Crust. et Insect.* (vol. i. p. 275.). A few naturalists with whom he was acquainted, then possessed considerable influence with the dominating party in the state; and Latreille happening to find, in his prison, what was then esteemed a rare insect (*Necrobia ruficollis*), found means of conveying it to them: this trivial incident drew their attention to his case; and they became so interested in his favour as to obtain his release. The names of his benefactors must not fail to be recorded: the chief of them were two naturalists of Bordeaux, MM. Bory de St. Vincent and Dorgelas; and they were aided in their exertions by a lawyer of the name of Martignac. In 1797, he incurred a risk of a similar kind, and was again indebted for his life to the zealous efforts of his friends, of whom he fortunately had many—some of them holding stations of importance.

When the state of public affairs again admitted of his residing once more in Paris, he received much kindness from M. Coquebert and his family: and through the efforts and recommendation of Lacepède, Cuvier, Geoffroy St. Hilaire, &c., he obtained employment in the Museum of Natural History; the duty being assigned him of arranging the insects in methodical order. About the same time he was nominated a corresponding member of the Institute. His literary and scientific occupations now became numerous and important, and he was marked as one of the most able naturalists of his day. Many of his insulated memoirs were inserted in Millin's *Magazin Encyclopédique;* others in the Annals and Memoirs of the Museum of Nat. Hist.; and not a few in the *Bulletin de la Société Philomathique.* They related chiefly to insects and the allied tribes of animals, but were by no means exclusively confined to them. A dissertation on the expedition of Suetonius Paulus into

Africa, another on the Atalantis of Plato, a third on
Egyptian chronology, &c., indicate that he was not entirely engrossed with natural history. But that constituted his chief occupation during the remainder of his
life; and the zeal with which he laboured may be inferred from the fact, that the memoirs and various works
which he published amount to between eighty and ninety.
His memoir on the sacred insects of the Egyptians,
and on the geographical distribution of insects, excited
much attention at the time when they appeared. His
" Précis des Caractères Génériques des Insectes," afterwards followed out in his great work on the genera
of insects, has always been regarded as a model of a
work of that kind. But it is impossible to enter into
any detail of his various merits in this place ; they were
such as eventually procured him a high reputation
throughout Europe ; and it is gratifying to reflect, that
they were fully appreciated in this country, as may be
seen by the eulogiums of Kirby, Leach, MacLeay, &c.
and other competent judges. His own country, never
slow to reward merit, elected him a member of the
Royal Academy of Sciences, as the successor of his
friend Olivier, in 1814. This was the first election
made by this institution, which was submitted to the
approbation of Louis XVIII. The monarch afterwards
conferred an additional proof of his esteem on Latreille,
by nominating him, in 1821, Chevalier of the Legion of
Honour.

In the latter years of his life, Latreille was Professor
of Zoology in the Veterinary and Royal School of Alfort; having been recommended to that station by his
predecessor. When the Entomological Society of France
was established, he was hailed as President, with as
general approbation as attended the appointment of our
own venerable Kirby on a like occasion in this country.
His death took place five or six years ago. He was buried
in Père la Chaise, where a handsome monument is erected
to his memory, as the " facilè princeps Entomologorum."
His remains were followed to the grave by most of the

scientific men in Paris, and a funeral oration pronounced over his tomb.—J. D.   His chief works are,—

1. *Précis* des Caractères Génériques des Insectes. Brives, 1796. 1 vol. 8vo.

2. *Salamandres, Hist. Nat. des.*   Paris, 1800. 1 vol. 8vo. with plates.

3. *Hist. Nat. des Crustacés* et des Insectes. Paris, 1802—05. 14 vols. 8vo., forming part of Sonnini's edition of Buffon.

4. *Hist. Nat. des Fourmis.* Paris, 1802. 1 vol. 8vo.

5. *Hist. Nat. des Reptiles* 4 vols. 12mo., forming part of Deterville's edition of Buffon.

6. *Genera Crustaceorum* et Insectorum, secundum Ordinem Naturalem in Familias disposita.   Paris, 1806—07. 4 vols. 8vo. pl. 16.

7. *Considérations* Générales sur l'Ordine Naturelle des Animaux composant les Classes des Crustacés, des Arachnides, et des Insectes.   Paris, 1810. 1 vol. 8vo.

LAURENTI, J. N. — *Erpetology.*

A physician of Vienna, and a very able erpetologist.

Specimen Medicum exhibens Synopsis Reptilium emendatum.   Vienna, 1768. 4to.

LEACH, DR. WILLIAM ELFORD. — *General Zoology.*

Dr. Leach will be long remembered in the zoological annals of this country, from having contributed more than any other to break down the strongholds of Linnæan nomenclature, and introduce the numerous improvements of the Continental nomenclature.   He was more especially attached, and was most conversant with, entomology ; and distinguished himself at a very early age, as enthusiastic in collecting British insects.   But he soon began to study them in detail ; and not only made himself thoroughly acquainted with the existing state of science, but visited Paris, and secured the friendship and

correspondence of Cuvier, Latreille, and the eminent
zoologists adorning the capital of France. Returning to
England, and possessing high interest (without which
his talents, in the eyes of the government, would have
availed him nothing), he succeeded in procuring the appointment of Assistant Zoologist to the British Museum.
It was in this situation, in the year 1817, that we first
had the gratification of making his acquaintance; this
acquaintance soon sprang up into great intimacy, which
continued uninterrupted to the sad event hereafter mentioned.   Dr. Leach found the zoological collections in a
most disgraceful state of decay and neglect.   It must
here be observed, that the salaries of this institution are
mostly so small, that in some cases they are insufficient
for the support of gentlemen.   The consequence is,
that, if they possess no private fortune, they are obliged
to become professional authors; and thus the duties of
the Museum are naturally neglected, the specimens unar-
ranged, and very often were suffered to decay in the name-
less crypts and vaults of Montague House: of these sub-
terraneous excavations there are so many, that they resemble the catacombs we have seen at Palermo, where one
is opened every day in the year, merely to deposit fresh
subjects for decay, and to ascertain how the process has
gone on during the last year.   In these visits we occasionally accompanied our friend; and very frequently assisted
him for hours in rummaging out and rescuing some of the
more valuable subjects from that oblivion to which they
were fast hastening.   It is not so now, we believe; but
such was the state of things when Dr. Leach took charge
of our national collection.   His predecessor had little or
nothing beyond his stipend.   His office, therefore, of
conservator, was nominal; for all his time was employed
in writing for the booksellers.   With Dr. Leach it was
far different.   A member of an ancient and independent
family in Devonshire, and possessed of a private fortune, his trifling salary barely served to purchase those
books and specimens necessary to his studies.   Dr.
Leach, in two years, did more to clear out these zoolo-

gical sepulchres, than his predecessor would have accomplished in a lifetime.   But he could not do all : he had
the work before him of five active zoologists, and he was
but — one.   Added to all this, he became absorbed in
entomology, and in many and various other similar and
interminable investigations.   Of a slight form, and delicate habit, it soon became evident to his friends he was
overworking himself.   His habits became irregular ;
he studied during the night, instead of the day ; and
although much attached to gymnastic exercises, they
were violent, uncertain, and not such as would operate
gently and effectually on a naturally nervous and irri-
table temperament.   One of these, for instance, was
leaping over the back of a stuffed zebra, which was
placed in the centre of a large room (now dismantled),
over which we have seen him vault with the lightness of
a harlequin.   In his conversation, he was particularly
animated, and even witty.   In his own apartments he
had two little rooms, — one containing a collection of
skulls, the other of bats ; these he designated as his Scul-
lery and his Battery.   He was so remarkably active, when
in health, that we have seen him leap up that long flight
of steps, constituting the present grand staircase, taking
three or four at a bound, and getting to the top, while
an ordinary person would have scarcely left the bottom.
It was, if we remember correctly, about the year 1821,
that we first remarked an evident change in his health :
he had become paler and thinner than usual — highly
nervous — and his eyes shone with that unnatural bright-
ness so indicative of the fearful affliction which soon
after became apparent to every one, and rendered the
care of his friends absolutely essential to his welfare.
This event, of course, terminated his public life at the
British Museum.   Removed from that exciting cause,
and retired in the country, his mind became more calm ;
and he soon sufficiently recovered to undertake a journey on the Continent, where he ever afterwards resided.
He finally died, we believe, at Rome ; and although his
physicians ever after prohibited him from close atten-

tion to his favourite pursuits, he still occasionally collected specimens for his friends. His name will be
long cherished by those who remember his warm, frank,
and generous disposition; and will ever rank high in the
science of this country, which, more than any other man,
he released from the thraldom of prejudice and bigotry.

Dr. Leach's works, unfortunately, are mostly in the
form of essays or papers, incorporated with scientific
periodicals or Transactions. We have enumerated the
greater part; but there are others in the *Zoological
Journal,* and some in the French periodicals.

1. *A Monograph* of the British Species of Meloe;
in the Linnæan Transactions, vol. xi.

2. *General Arrangement* of the Classes Crustacea,
Myriopoda, and Arachnides; in the Linn. Trans.
vol. xi.

3. *On the Notonectides,* in the Linn.Trans. vol. xii

4. *Proboscidious Insects,* on the Genera and Species
of ; in the Wernerian Transactions for 1817.

5. *Malacostrata Podopthalmos Britannicæ.* 4to.
London, 1815, 1816. The figures by Sowerby. A
beautiful work; but never completed, as it terminated at the eighth number.

6. *Zoological Miscellany,* being Descriptions of
new or interesting Animals, illustrated with coloured
Figures by R. P. Nodder. London, 1815. 3 vols.
royal 8vo. The figures, excepting those of the birds
and quadrupeds, are very good. Besides these, Dr.
Leach considerably assisted Mr. Samouelle in the materials for his Compendium : he also wrote a short
account of the animals discovered by Bowdich and
Crouch during their respective travels.

LE CONTE, JOHN. — *Entomology, &c.*

A major in the army of the United States; an acute
and most liberal naturalist. The only distinct publication which bears his name, and of which he is certainly
the real author, is

Histoire Générale, et Iconographie, des Lépidoptères ou des Chenilles de l'Amérique Septentrionale.
Paris, published in conjunction with Boisduval. A
very useful work, but far inferior to that of Abbots.

LEISLER. — *Ornithology.*

A German ornithologist, known as having written a
Supplement to Bechstein's Birds of Germany, printed in
1812 and 1813.

LEPELLETIER DE ST. FARGEAU, A. — *Entomology.*

An eminent entomologist, and a most accurate describer, well known for his excellent monograph of the
French *Chrysidæ.*

1. *Monographie des Chrysis* des Environs de Paris :
published in the Annals du Muséum d'Histoire
Naturelle.

2. *Mémoire sur les Araignées.* — Bulletin de la
Société Philomathique for 1813.   The only separate
work he has yet published is the following : —

3. *Monographia Tenthredinetarum,* Synonymia ex-
tricata.   Paris, 1823. 1 vol. 8vo.

LESKE, N. G. — *General Zoology.*

Born in 1752 ; was made Professor at Leipsig, and
afterwards at Marpurg : he died in 1786.

Museum Leskeanum.   Leipsig, 1789. 2 vols. 8vo.
coloured plates.   There is another edition, somewhat
enlarged, by Klein.   Leips. 1778. 4to.

LESSON, R. P. — *General Zoology.*

A zealous naturalist, and most laborious writer and
compiler.   Both he and M. Garnot were attached, as
surgeons and zoologists, to the Coquille, on her voyage
to the Pacific, under the command of captain Duperry.

R

On their return to France, they published the zoological results of their expedition. Subsequently M. Lesson produced, in rapid succession, a number of works, some of which, as may be expected, bear the marks of great haste and oversight.

1. *Voyage autour du Monde,* exécuté par Ordre du Roi, sur la Corvette La Coquille, pendant les Années,1822—1825. Part. Zoologie, par Lesson et Garnot, Médecins de la Marine Royale.— Crustacés et Insectes, par F. E. Guérin. 2 vols. 4to., and an Atlas of 150 coloured plates in folio. Paris, 1827,&c. The engraving and colouring of these plates are done with the greatest care ; but it is obvious that many of the figures have been taken from poor, and even rude sketches, finished up by the Paris artists.

2. *Manuel de Mammalogie.* Paris, 1827. 1 vol.

3. *Manuel d'Ornithologie.* Paris, 1820. 2 vols. 12mo.

4. *Manuel de l'Histoire des Mollusques,* et des leurs Coquilles. Paris, 1829. 2 vols. 12mo.

5. *Histoire Naturelle des Oiseaux Mouches.* Paris, 1829. 1 vol. royal 8vo., complete in 17 numbers, with 86 highly coloured plates.

6. *Hist. Nat. des Colibres*, suivie d'un Supplément à l'Histoire Naturelle des Oiseaux Mouches. 13 numbers, royal 8vo. with 66 plates. Paris, 1831.

7. *Les Trochilidés,* ou les Colibres et les Oiseaux Mouches. 14 numbers, with 70 plates. Paris, 1832—34.

8. *Centurie Zoologique,* ou Choix d'Animaux rares, nouveaux, ou imparfaitement connus. 16 numbers, royal 8vo. with 80 plates. Paris, 1831—32.

9. *Illustration de Zoologie,* ou Choix de Figures, peintes d'après Nature, des Espèces inédites et rares d'Animaux, recemment découverts. 10 numbers, with 30 plates. Paris, 1832—34. The above five works are printed uniformly. The figures, which are all designed and engraved by the Paris artists, are perfectly characteristic of their peculiar style of

representing subjects of natural history; that is, well
engraved, and beautifully coloured, but destitute of
effect, chasteness of drawing, or of good perspective.
This is no fault, of course, of the author, whose descriptions are for the most part exact; although
many of the subjects, he supposes to be new, had been
previously described.

10. *Traité d'Ornithologie,* ou Description des
Oiseaux réunis dans les Principales Collections de
France.   Paris, 1831.   2 vols. 8vo. with 119 plates.

11. *Histoire Naturelle des Oiseaux de Paradis,*
des Sericules, et des Epimaques.   1 vol. royal 4to.
45 plates.   Paris, 1833.

12. *Histoire Naturelle,* générale et particulière,
de tous les Animaux rares et nouveau, découverts par
les Naturalists et les Voyageurs dépuis la Mort de
Buffon; formant le Complément indispensable des
Œuvres de Buffon.   10 vols. 8vo. with an Atlas of
20 plates.

13. *Journal Pittoresque d'un Voyage autour du
Monde.*   To form 3 vols. in 8vo , with 30 plates by
the best artists. — N. B. Of the last three works, we
have only seen the prospectus, and know not whether
they have been completed.   It seems altogether inconceivable, how even the manual labour of so many
works could have been done, and got through the press,
in the short space of four years!

LE SUEUR, CH. A.— *General Zoology.*

It is hardly necessary, in this place, to repeat the high
estimation in which we hold the talents of this inimitable painter, accomplished naturalist, and accurate describer.   He was, says Cuvier, one of the draftsmen
who accompanied Péron in the discovery ships of
Baudin to the Australian seas, and was his most effi-
cient and zealous co-operator in zoological researches.
He has published some zoological observations in the
Bullétin des Sciences, and the prospectus of a great

work on the *Medusæ*, accompanied by specimens of the
plates. On what account this was relinquished, and
why this accomplished man emigrated from his country,
to settle in America, we know not: certain, however, it
is, that he left behind him no one, in France, who was
qualified to fill his place, or whose delineations for a
moment can be compared to his own. The fostering
protection of the French government does not appear
to have been extended, in this instance, to one who had
the highest claims upon its liberality. Inferior and
commonplace artists are attached to the establishment
of the French Museum, while the Raffaelle of zoological painters was suffered to emigrate, and pursued his
professional career as a private teacher in Philadelphia,
where, we believe, he now is. This eulogium is not dic-
tated by partial friendship. We know Le Sueur only by
his works,—his outline figures, engraved by himself, in
the American Transactions; and the many valuable
papers on fish, *Crustacea*, and *Mollusca*, from his pen,
which these plates illustrate. It is deeply to be regretted, that his works are so scattered, in collections of
papers hardly ever seen in Europe; and that no one
volume will hereafter point out the matchless excellence
of Le Sueur.

LEUKARD, F. S.—*Malacology.*

1. *The Mollusca* in Rüppell's Atlas to the Zoology
of Northern Africa. (See RÜPPELL.)
2. *Zoological Fragments.* Helmstadt, 1819.

LEVER, SIR ASHTON, BART.— *General Zoology.*

The name of this amateur naturalist, although not
attached to any separate work as an author, will long be
remembered in the zoological annals of Britain, from his
having expended an immense fortune in the formation
of what was once the Leverian Museum. Sir Ashton
was the son of sir D'Arcy Lever, an ancient Lancashire

family, long settled at Alkington, near Manchester, a town, in fact, which contains, within its vicinity, more collections of art and nature than any other of similar extent in the British dominions. Sir Ashton's passion for collecting exceeded all bounds of prudence : every subject of zoology or mineralogy he did not possess, was to be purchased, cost what it might. On this plan, it may be readily conceived that every thing new or valuable, intended for sale, was first offered to him; and his museum thus began to rival, in the department of zoology, and ultimately to excel, that of sir Hans Sloane, already in the British Museum. The most princely fortune, however, could not bear up against such continued heavy demands, and sir Ashton's affairs became ultimately embarrassed. His collection, naturally, had now assumed such an extent, and was so highly thought of, that his friends suggested to him the expediency of removing it to London, where the funds arising from its exhibition might produce a sort of annual interest upon its original cost; or, at least, furnish the means of purchasing fresh additions. Sir Ashton, accordingly, removed the whole to London, and about 1775 it was opened for public inspection, in Saville House, Leicester Square,—a building still existing as a place for exhibition, although reduced, in these days, to one half its original dimensions. For the first few years, the result completely answered the expectations of the liberal proprietor ; and enabled him not only to mix in the first circles of intellectual society, on an equal footing as regarded his establishment, but also to add still more to his noble collection,—which now, in regard to preserved animals, eclipsed the British Museum. But the public requires the stimulus of novelty to insure its attention ; and after a few years, the annual amount of receipts diminished so much, that sir Ashton's creditors became clamorous, and he found himself compelled, in the year 1785, to dispose of the whole by way of lottery. It would be curious, at this time, to ascertain the particular " scheme" by which this was effected; and the col-

lectors of ancient tracts and forgotten hand-bills would
do well to ferret out the particulars : we only remember
the old naturalists of our early days mentioning that there
was but a single prize — that of the whole museum ;
and that it was gained by Mr. Parkinson, who thus, from
a dentist, was compelled to become a naturalist. A
large and very appropriate building was soon after
erected for its reception, on the Surrey side of Blackfriars Bridge. This was a most injudicious site, since
it was completely out of the mighty stream of human
beings which never ceases to flow through the centre
of the metropolis. We believe sir Ashton did not long
survive the relinquishment of what had cost him so much
to acquire. Our father, who knew him well, used to
describe him as a high-bred, but most agreeable man,
with a cultivated mind, but by no means inclined to dry
scientific research, or prone to investigate the vast materials he had thus heaped together. His love for natural history, in fact, centered in the possession of what
was beautiful, curious, or rare, — the gratification of the
eye, more than the enlargement of the mind. It was
one of sir Ashton's peculiarities, in his latter days, of
always carrying and drinking his own wine, at whatever
dinner party he attended. People in general thought
this very odd ; but considered the worthy baronet was
medically obliged to follow this rule, and never inquired
further. It was then the custom for gentlemen of a
certain rank to carry their own man-servant to the house
they dined at, to attend them at table ; and sir Ashton's
valet regularly filled his master's glass (and that not a
small one) from his own bottle ; the liquor seemed very
dark Madeira, but his intimate friends soon discovered
it was always a light *brandy*. In those days, when
drunkenness was gentlemanly, we know not how often
these potations were repeated ; but they must inevitably
have shortened life. The Leverian Museum — now removed beyond the immediate ken of the " discerning
public," — did not answer the expectations of its new
possessor. Yet it was still one of the sights for holyday

folk, and country cousins ; and more than once we remember passing a morning of delight in its large and
handsome circular gallery, well lighted from above, and
resplendent with innumerable cases of birds. Nevertheless, its decay was inseparable from the spot it had
now been fixed upon ; and, about the year 1805, it was
finally disposed of by public auction : the sale extending
to between 28 and 30 days. It attracted naturalists
from all the courts of Europe, and realised a very handsome sum. Mr. Parkinson, or his son, subsequently
became appointed British consul at Pernambuco, in
Brazil.

LEWIN, WILLIAM. — *Zoological Artist.*

Lewin was the best zoological painter, and one of the
most practical naturalists, of his day. He was much
patronised by the great duchess of Portland ; and received attention and constant encouragement from Dr.
Fordyce, the late John T. Swainson, and other eminent
men of that time. He generally painted his subjects
in body colours, upon vellum : his style was bold, and
his colouring powerful, without, in general, being highly
finished. His etchings bear the same character ; they appear to have been done with too much rapidity, to allow
of that precision and clearness in the details so necessary
in subjects of this nature. His birds are always easy,
full of animation, and very correct in their proportions ;
yet, in his attitudes, he sometimes overstepped nature.
His drawings of shells have never been published, but
a very extensive and valuable collection, expressly executed by him for the late J. T. Swainson, Esq., are
now in the possession of his daughter, Mrs. Willis, of
Liverpool, and evince the artist's high talents in this
department. We have seen some exquisitely finished
drawings of Lewin upon vellum, of British insects,
which might be compared to those of Van Huysum.
Lewin left two sons, both of whom followed the profession of their father. One went as an artist to New

South Wales ; but died poor, and left a widow who
returned to England : the other, I believe, still follows
his profession in London, with more honour than profit.
Cannot the nation afford to have a *single* artist attached
to the British Museum ?

The Birds of Great Britain, systematically arranged, and accurately engraved and painted from
Nature, with Descriptions, including the Natural History of each Bird.　London, 1796—1801.　8 vols.
4to.　This edition contains 267 plates of birds, independent of 56 figures of their eggs.

LEWIN, J. W.— *Ornithology and Entomology.*

Born in ——— ; died at Sydney, New South Wales,
1821.

1. *The Birds of New South Wales,* and their Natural
History, collected, engraved, and faithfully painted
after Nature.　4to.　26 coloured plates.　There are
two editions of this work, although, unfortunately, it
was never extended beyond one thin volume.　The
figures, although slightly and sometimes even rather
coarsely etched, are yet admirable, and full of truth
and nature ; while several valuable observations on
the habits and economy of the birds themselves, are
scattered in the text.

2. *Lepidopterous Insects* of New South Wales,
Natural History of the.　Lond. 1805.　1 vol. 4to.
with 18 col. plates.

LICHTENSTEIN, H.— *General Zoology.*

This celebrated traveller and learned naturalist, is
the chief Professor, and we believe director, of the
Royal Museum at Berlin.

1. *Travels in the Cape of Good Hope.*　Berlin,
1811. 2 vols. 8vo.　This work has been translated
into English, in 1 vol. 4to., and we have frequently

quoted it for many valuable notices on the animals met with by the author.

2. *Darstellung* neuer oder wenig bekannter Sangthiere in Abbildungen und Bechreibungen nach den Originalen des Zoologischen Museums der Universität zu Berlin. Berlin, 1827—1833. 8 parts, folio, with 42 coloured plates.

His other works chiefly consist of essays on the antilopes, the Brazilian quadrupeds noticed by Marcgrave, &c., which are inserted in the Berlin Transactions: another on the eye-like spots (or *crepitaculum* of Guilding) on the wings of the *Locustæ*, is in the Linn. Trans. iv. p. 51.; the sixth volume of which contains also his dissertation on the genera *Mantis* and *Phasma*. M. Lichtenstein visited England at the sale of Bullock's Museum, and speaks our language remarkably well. His manners are particularly agreeable. He has a very fine taste in music, and is himself an accomplished performer. The portrait affixed to his travels is a likeness, but certainly not a flattering one.

LINK, J. H. — *Malacology.*

A physician of Leipzig, born in 1674, died in 1734, whose volume is still of value.

De Stellis Marinis, Liber singularis ; edited by Ch. Gab. Fischer. Leips. 1733. folio.

LINNÆUS.

This great founder of systematic natural history, whose name is familiar to all the civilised world, was the eldest son of Nils or Nicholas Linnæus and Christina Broderson, and born at Rashalt, in the province of Swaland (Sweden), on the 24th of May, 1707. His father was pastor of the village just named, and was the first member of his family that had received a liberal education, — his parents being poor peasants. Charles was designed by his father to succeed him in the pastoral office ; and, at the age of seven, he was sent to a school in the neighbouring town of Wexio. Here, however, his progress by no means came up to his father's wishes, for he showed no aptitude for classical learning, but made his escape, whenever he possibly could, to the fields, in order to search for plants and insects. Even when removed to the upper school, or gymnasium as it was then called, the same propensities prevailed, and were daily acquiring additional strength; insomuch that his father at last despaired of his qualifying himself for the study of divinity, and resolved to bind him apprentice to a shoemaker. Happily for the young naturalist, and the cause of science, a benevolent physician, Dr. Rothman, befriended him at this crisis, and took him to reside with him for a year, till he should complete his course at the gymnasium. It was by the advice of this valuable friend, that it was finally resolved that young Linnæus should study medicine. This enabled him to give full scope to his passion for plants; and he eagerly entered upon the study of Tournefort's Institutions, the works of Tilland, Manson, Bromellius, and others which are now almost forgotten.

With a certificate which contained not a single word in favour of his scholarship, in 1797, Linnæus repaired to the university of Lund, into which he obtained admission through the interest of one of his old teachers, Gabriel Hök. He was placed under the care of Kilian Stobacus, Professor of Natural History, whose favour he soon gained by his diligence ; and he enjoyed many comforts in consequence. After recovering from a severe illness, produced, he himself affirmed, by the bite of a singular worm (*Furia infernalis*), he visited his parents, and former benefactor, Dr. Rothman ; by the latter of whom he was persuaded to go to the university of Upsal, in the prospect of enjoying greater opportunities of improvement. Having set out with so small a provision as 8*l.*, — the whole that his parents could afford him, — it was not very long before his pecuniary

means became exhausted; and he is said to have been
so reduced as to be under the necessity of repairing the
cast-off shoes of his companions for his own use. He
was ultimately relieved from his difficulties by professor
Rudbeck and Dr. Celsius; the latter of whom received
him into his house, and conferred on him other favours,
which he was enabled, in some measure, to repay by assisting Celsius in the preparation of his *Hiero-Botanicon.*
Professor Rudbeck afterwards appointed him tutor to
his children, and employed him as assistant in his pro-
fessorial duties. About this time he laid the foundation
of many works, afterwards published, and even in part
executed them: he also acquired many friends, who
subsequently co-operated with him; among others, he
speaks of Artedi with much regard.

In the spring of 1732, he set out to make a scientific
tour through Lapland, having been chosen for this purpose by the Royal Academy of Upsal. This arduous
expedition he accomplished in a highly profitable manner; returning with numerous novel and important contributions to natural science. He now wished to lecture
publicly at Upsal; but his design was frustrated by the
jealousy of the professors, who alleged that he was le-
gally disqualified, in consequence of not yet having
taken his degree. He then retired to Fahlun, a town
situate in a celebrated mining district of Dalecarlia,
where he was appointed to superintend some young
men in examining the mineralogy and natural history
of the country. It was here that he became enamoured
of the daughter of Dr. Moræus, a wealthy physician,
and sought her in marriage; but the father's consent
could not be obtained till after a probation of three
years, and his having procured his degree of M.D.
This he obtained from the university of Harderwick,
along with flattering testimonials of his high abilities
and attainments. On returning through Holland, he
became acquainted with many of the most distinguished
Dutch literati, in particular, Gronovius, professor Bur-
mann, and Boerhaave. The former of these induced

him to publish his *Systema Naturæ,* which was comprised in 14 folio pages ; the second engaged him to
assist in a work on the plants of Ceylon ; and by the
latter he was introduced to Dr. George Clifford, a wealthy
cultivator of plants, by whom he was most liberally
treated, and commissioned to visit England for the purpose of procuring botanical novelties. His reception by
English botanists, of whom the principal at that time
were sir Hans Sloane and Dillenius, was any thing but
cordial,—they, doubtless, regarding him as a rash inno-
vator in science ; but they ultimately acknowledged his
merits. On a shortly subsequent occasion, he visited
Paris likewise, where he met with a more favourable
reception, and formed a lasting friendship with Bernard
de Jussieu.   Among other honours then conferred on
him, he was elected a corresponding member of the
Royal Academy of Sciences.

He returned to his native country in the autumn of
1738, having been absent three years and a half. After
numerous mortifications and disappointments, he began
to practise medicine in Stockholm,—at first with little
success, but in time he became highly popular.   By
the influence of count Tessier, he was appointed to the
office of lecturer to the School of Mines, and soon afterwards physician to the Admiralty.   To these honours
he added, in 1739, those of physician to the king, and
president of the Academy recently formed in Stockholm.
In this year also he was united to Sarah Elizabeth, the
daughter of Dr. Moræus, to whom he had been so long
contracted.   Two years later he reached the summit of
his ambition, by obtaining a chair in the university of
Upsal, — at first that of medicine ; but he afterwards
effected a change, by which the whole department of
natural history was placed under his superintendence.
This chair he occupied for thirty-seven years, surrounded by pupils, many of whom were attracted from
great distances by his fame, — his reputation daily extending,— enjoying the favour of his sovereign, and the
affection and respect of most of the eminent men of

Europe,—his system of natural history (in particular, the sexual system of plants), gradually overcoming all opposition,—and his native country, through his means, regarded by all as the quarter from which light was to emanate to guide them to the knowledge of nature. Medals were struck in his honour; invitations of the most flattering kind sent to induce him to come and reside at foreign courts; most of the learned Academies of Europe elected him an honorary member; and, in 1761, he was presented by his sovereign with letters of nobility. His circumstances being now affluent, he lived in corresponding style; delighting to assemble round him his favourite pupils at his villa of Harmanby, and blend his instructions familiarly with the hospitalities of social life. Many of his scholars, to whom he had communicated, as he had a singular facility in doing, a portion of his own enthusiasm, were sent to foreign countries to collect objects in natural history, and he was continually receiving fresh supplies. This agreeable state of things continued for many years, and was only interrupted by the decline of his constitution. In 1750, he had a severe attack of rheumatism or gout, which he relieved by eating strawberries; and it seems to have recurred at intervals ever after. In May, 1774, he had an attack of apoplexy: two years afterwards, this was succeeded by another; and an accumulation of other disorders cut him off on the 10th of January, 1778, at the age of 71.

He left one son and four daughters. The former was appointed assistant and successor to his father in the professorship, at the early age of 21. He afterwards travelled through various parts of Europe; and died of apoplexy on his return to Upsal, in the 42d year of his age.

The museum of the elder Linnæus, as is well known, was purchased by the late sir J. E. Smith, and is now in possession of the Linnæan Society of London. The enumeration of his works would fill many pages.—J. D.

LISTER, MARTIN, DR.— *Malacology and Entomology.*

Celebrated as a malacologist, and a diligent cultivator of natural history in other departments. He was born at Radcliffe, in the county of Buckingham, in 1638. His family, for some generations, had followed the medical profession; his grand-uncle, sir Martin Lister, having been physician in ordinary to Charles I. His education was finished at St. John's College, Cambridge. He then visited the Continent, according to the common practice among gentlemen of his profession at that period; and on his return, settled at York. Notwithstanding his professional labours, which are said to have been unremitted, he devoted much time to natural history; and we find him corresponding from that place, on subjects connected therewith, with sir Hans Sloane, Ray, &c. Between him and the latter a very intimate friendship existed; and after the death of Willughby, he invited Ray to come and reside with him at York. The study of antiquities, at this time, divided his attention with that of natural history; and in prosecution of both, he travelled frequently about the northern counties of England. These propensities brought him into connection with Mr. Lloyd, conservator of the Ashmolean Museum at Oxford; and that institution became indebted to him for many valuable objects both in antiquities and natural history. His manuscripts, also, relating to these subjects, were transmitted to Mr. Lloyd, by whom they were sent to the Royal Society of London. Of this society, Lister himself was soon after admitted a member.

Dr. Lister left York in the year 1684, and removed to London. About the same time he took the degree of Doctor at Oxford, and not long afterwards was elected a Fellow of the Royal College of Physicians. When the duke of Portland was appointed ambassador to the court of France by king William, in 1698, Lister accompanied him in his professional capacity; and on his return, published an account of his "Journey to Paris." This journey was attended with no very important incidents; and in want of these, the author gave a very minute detail of ordinary occurrences, and many curious observations in natural history. Notwithstanding the value which now attaches to some of these details in a historical point of view, they were not highly appreciated at the time; and a Dr. William King attempted to turn them to ridicule in a parody entitled a "Journey to London." At the age of seventy-one, he was appointed physician in ordinary to queen Anne; but this honour he enjoyed for a very short time,—his death having taken place on the 2d of February, 1711.

His medical writings are pretty numerous, but are not of great value, as he was too prone to indulge in hypotheses, and had more respect for the *dictum* of the ancients than the evidence of observation. The very reverse of this, however, may be affirmed regarding his works on natural history, which are characterised by accurate observation, great knowledge of comparative anatomy, and, in general, just notions of the natural affinities of animals. His various works on shells have laid the foundation of all precise knowledge on this subject. It was of great use to Linnæus, who eulogises it as the richest (*ditissimus*) work on the subject which had appeared up to his time. The first edition of this valuable publication is now rarely met with; but subsequent editions may be procured without difficulty. Next to this work, his " Treatise on Spiders " is perhaps the most valuable; it was originally written in Latin, but has been translated both into German and English.— J. D.

1. *Historia Animalium Anglicæ,* Tractibus duo de Cochleis tum Terrestribus tum Fluviatilibus, et de Cochleis Marinis, impr. cum ejus Tractatu de Araneis. Lond. 1678. small 4to. plates.

2. *Historiæ sive Synopsis Methodicæ Conchyliorum* et Tabularum Anatomicarum. Lond. 1685—1693. 1 vol. folio. 1057 plates, upon 468 leaves.

Ditto. Editio altera, recensuit et Indicibus auxit Gul. Huddesford. Oxon. 1770. Plates, 1059 shells,

22 Anatomicæ. A third edition was published in
1823, with a scientific Index by Mr. Dillwyn. The
first edition, Mr. Wood observes, was published in
London in 1685 : the copies, which are rarely found
complete, ought to contain 1057 plates, on 468 leaves,
independently of the 22 plates in the Appendix. In
the Oxford edition, there are 1059 plates of shells,
besides the 22 anatomical subjects; also six pages of
Lister's notes and observations, not to be found in
the first edition ; and two very imperfect indexes at
the end of the volume.

LORD, ——.— *Ornithology.*

History of British Birds. London, 1791. folio.
pp. and pl. 96. It is difficult to conceive figures
worse executed than those contained in this book, now
seldom seen.

LYONNET, PETER.— *Entomology.*

Celebrated for having devoted several years to the investigation of a single insect ! He was perpetual secretary to the United Provinces.

Pierre Lyonnet, alike distinguished as a naturalist, an
anatomist, and an engraver, was born on the 21st of
July, 1707, at Maestricht. His father, who was pastor
of the French church at Heusden, destined his son for the
same sacred calling ; but when the latter came of age to
judge for himself, he preferred the study of the law,
which he accordingly prosecuted at Utrecht. Having
an extraordinary aptitude for acquiring languages, he
made himself master, at an early age, of most of the
modern European tongues ; and his proficiency in these,
as well as in classical learning, procured for him the
appointment of sworn translator to the States-General.
The abundant leisure which the duties of this office left
on his hands, he employed in collecting and drawing the
insects in the neighbourhood of the Hague, in translating

Lesser's *Théologie des Insectes* (to which he added numerous valuable notes), and in drawing and engraving a portion of the plates for Trembley's famous work on the freshwater polypus. His success in this first attempt to use the graver so encouraged him, that he resolved further to employ it in a work of his own; and the fruit of this resolution was his famous treatise on the anatomy of the caterpillar of the *Cossus*. This work first appeared in 1760, forming a quarto volume of upwards of 600 pages, with 18 plates; and it has ever been considered, for minute accuracy of delineation, and incredible delicacy of execution, as a perfect model of a work of this nature. It formed part of his design to illustrate in a similar manner the anatomy of the chrysalis, and perfect moth; but his labours were interrupted by an accident which impaired his eye-sight. A portion of the plates prepared with this view, and others of a miscellaneous character, appeared in a quarto volume a few years ago; and it is reported that other papers of value are still in possession of Lyonnet's descendants. He died on the 10th of January, 1789, at the age of eighty-two years. His collection of shells, which was extensive and curious, was sold at the Hague in 1796. —J. D.

Traité Anatomique de la Chenille du Saule. La Haye, 1762. 4to. with plates engraved by the author. A work, observes Cuvier, "which is at once the masterpiece of engraving and anatomy,"—and well it should be.

MACCARI, PIERRE. — *Entomology.*

Member of the Medical Society of Marseilles.
Mémoire sur le Scorpion qui se trouve sur la Montagne de Cette, &c. 1810. 1 vol. 8vo.

MACRI, XAVIER.

New Observations on the Natural History of the Ancients (in Italian). Naples, 1778. 8vo.

MACLEAY, WILLIAM SHARPE. — *General Zoology.*

The celebrated author of the *Horæ Entomologicæ;* being the first philosophic exposition of the circular system of affinities, &c.*

1. *Horæ Entomologicæ;* or, Essays on the Annulose Animals. 1 vol. 8vo.   London, 1819—1821.

2. *Annulosa Javanica*, No. 1. with a plate. London, 1825. 4to.   The departure of the accomplished author for Cuba, the following year, prevented this valuable work from being continued.   Mr. Macleay has since settled in New South Wales, where his family possesses considerable property.

MACQUART, J. — *Entomology.*

1. *Insectes Diptères* du Nord de la France; with plates representing their wings; published in the "Transactions of the Royal Society of Lille," which form 7 vols. 8vo. with plates. Lille, 1826—29. (*Cuv.*)

2. Histoire Nat. des Insectes (Diptères).   2 vols. 8vo.   Paris, 1834—35.

MANNERHEIM, C. G. — *Entomology.*

An able entomologist, councellor to the emperor of Russia.

1. *Eucnemis Insectorum Genus.* 1 vol. 8vo.; with two plates.   Petrop. 1823.

2. Précis d'un nouvel Arrangement de la Famille des Brachelytres, in the Petersb. Trans. for 1830.

3. *Forty new Species* of Scarabæides from Brazil, Description of; with plates in 4to.

* See Classification of Animals, p. 201.

MARCGRAVE, GEORGE.— *General Zoology.*

George Marcgrave, who has been designated the Father of Brazilian Zoology, was born in the year 1610, at Liebstadt, in Meissen, Upper Saxony. He was brought up to the medical profession; and having early distinguished himself by his knowledge of mathematics and natural philosophy, he was selected by the eminent Pison to accompany him in the suite of prince Maurice of Nassau-Siegen (see under MAURICE) in the Dutch expedition, which set off in 1636 to conquer the Portuguese settlements in South America. The young naturalist speedily recommended himself to the kind offices of the prince himself, who engaged him in his more immediate service, employed him in the prosecution of science in the newly conquered provinces, and left nothing undone to supply him with a safe conduct and every possible convenience. Under these favourable circumstances, he was, for about six years, engaged in these exploratory journeys; and, on the return of prince Maurice to Holland, was excited by his thirst of additional acquirements and opportunities, to undertake a voyage to the coast of Guinea; where he died, a victim to the climate, in the year 1644.

His manuscripts naturally fell into the hands of the prince, who entrusted those which bore on astronomy to professor J. Golius; and to De Laet, those which had reference to natural history. The plan prescribed was, that the writings of Marcgrave should be published in the same volume with the somewhat similar labours of his early friend and patron, Pison, whilst they were at the same time to be kept quite distinct. Of his works on the former subject, a fragment only, we believe, has seen the light; but those on zoology, through the energy and ingenuity of De Laet, were much more fortunate. The task, however, was by no means an easy one. The work, as left by its author, was far from being in a finished state; and what added greatly

s 2

to the labour, was the circumstance that Macgrave, ap-
prehensive, according to De Laet, that if any misfortune overtook him, others would furtively appropriate
his materials, employed, in writing a great part of his
observations, and especially the most important, a character which was altogether arbitrary.   Recourse was
therefore necessary to an alphabet he had carefully con-
cealed, that his cipher might be discovered and read.
De Laet readily undertook and mastered this part of the
task, added explanatory notes, and superintended the
publication of the two works,— that of Pison, under
the title of *Medicina Braziliensis,* and of Marcgrave, on
its natural history.   Our author here introduces to
notice an immense number of new plants, and supplies
the names he had received from the natives : the majority of his descriptions have subsequently been found
to be accurate and satisfactory ; and the woodcuts, taken
from his drawings of plants and animals, are often respectable.   This work was published in the year 1648;
and must have been got up, probably to meet the keen
curiosity which was prevalent at the time, in great
speed—not to say hurry.   Its execution was, at all
events, decidedly unsatisfactory to the surviving author,
Pison, who, ten years later, brought out another edition,
entitled, *The Natural History and Medicine of both
Indies ;* in which greater prominency is given to his own
labours ; incorporating therewith those of Marcgrave,
and adding those of Bontius, now first published, on
Batavia.   Here Pison freely retrenches the work of
Marcgrave, wherever he considers his statements of
minor importance ; and he has done this, not without
the reproach of having availed himself of many without
due acknowledgment.   In this volume, however, he introduces, separately, a short tract of Marcgrave's on the
topography and meteorology of Brazil, with observations upon a solar eclipse which had occurred there :
and by De Laet, we are supplied with the title of a
greater work, which our indefatigable naturalist had
contemplated rather than completed, treating especially

on the astronomy, geography, and geology of the Brazils.

Thus was cut off, at the age of 34, a man of great talents and acquirements; which, for a few years, were busily and variously employed in distant and dangerous climes, in enriching the archives of science, and adding to the stock of human knowledge, thereby rearing a reputation he lived not to enjoy.—J. D.

Historiæ Rerum Naturalium Braziliæ. Libri VIII. Amst. 1648. folio.

MARIUS, J.— *Mammalogy.*

Traité du Castor. Paris, 1746. 12mo.' pp. 280. pl. 3.

MARSHAM, THOMAS.— *Entomology.*

A zealous Linnæan entomologist and collector; for many years treasurer of the Linnæan Society.

Entomologia Britannica, sistens Insecta Britanniæ indigena, secundum Methodum Linnæanum disposita. Tom. 1. Coleoptera. London, 1802. No other volume was published. There is a valuable paper by this author on the Australian genus *Notoclea* (or *Paropsis*), illustrated by many figures, in the Linn. Trans. vol. ix.

MARTIN, ST. ANGE, AND GUÉRIN.

Traité Elémentaire d'Histoire Naturelle, comprenant l'Organisation, les Caractères, et la Classification des Végétaux et des Animaux. Paris, 1833. 2 vols. 8vo.

MARTINI, F. H.— *Conchology.*

Frederick Henry Guillaume Martini, now chiefly known for his works on conchology, was born on the

s 3

31st of August, 1729, at Ohrdruf, in the duchy of
Gotha. He was sent to Jena to study theology; but his
health was delicate, and he ultimately became a physician. He obtained his degree of Doctor of Medicine
in 1757, and went to practise at Artern, near Mansfield,
where he devoted all his leisure hours to the study of
nature, to which he had always been strongly inclined.
At the end of four years he went to Berlin, and established himself there. By his exertions, a Natural History Society was instituted in that city, in 1773; and
he was elected perpetual secretary. He died five years
after that event; his death, it is said, having been much
accelerated by intense application to study.

This statement may be regarded as receiving some
confirmation from the voluminous works which he published in the space of fourteen years. Besides numerous
translations and separate articles, we have his great
System of Conchology, ultimately extending to eleven
quarto volumes, several of which were the result of his
own labour; a *Dictionary of Natural History*, in
eleven volumes, &c.—J. D.

> *Neues Systematisches Conchylien Cabinet.* Nu-
> remberg, 1769—1800. 12 vols. 4to.; the last con-
> taining a general Index. The plates are numbered
> in two series: the first contains 193, and ends with
> vol. 5.; the second, 213, and terminates with vol.
> 11. Considering the industry and perseverance be-
> stowed in gathering the materials for this great work,
> we cannot but regret they have been so ill digested and
> arranged: the nomenclature can rarely be depended
> upon; and the figures, in the early volumes, are so
> inaccurate, that many can scarcely be recognised.
> Nevertheless, the work is indispensable to every con-
> chologist who studies, from the vast number of species
> it contains, and being almost the only one of its kind.

Mauduit, R. J. E. — *Ornithology.*

Author of the ornithological part of the *Encyclopédie Methodique,* a work subsequently revised and augmented by Vieillot.

Maurice, or Mauritz, Prince or Count of Nassau-Siegen. — *General Zoology.*

The annals of science contain no name more truly deserving record than that of the learned and chivalrous Prince Maurice of Nassau, the conqueror and natural historian of Brazil.   If one of the elements of greatness consists in the power of excelling in every thing that the mind undertakes, more especially in pursuits diametrically opposed to each other, history can scarcely furnish a more conspicuous instance than that of the extraordinary man whose career we shall now briefly sketch.   " In every circle one may meet with men of prodigious energy and of indefatigable zeal : but they are such as can exist only exteriorly, or in action : rest, when it must be taken, is, with them, a cessation of intellectual life, not another and a graceful mode of it."   " We might find," continues the same beautiful and philosophical writer, " plenty of great minds, if we could but relinquish in our definition this peculiar characteristic, a tranquil taste, and the capability of repose *,"— united, we may add, to the achievement of great and heroic enterprizes. If, this, as we believe, is one of the tests of true greatness, few men deserve that distinction more than him of whom we shall now speak.

Prince John Maurice, of Nassau-Siegen, was born in 1604.   He was grandson of John, count of Nassau, chief of the branch of Dellenburg, and must not be confounded with his namesake of the house of Orange, so celebrated in those wars which released Holland from the Spanish

* *Saturday Evening,* by the author of the *Natural History of Enthusiasms.* Fifth edition, post 8vo. p. 254.
s 4

yoke. From early youth he manifested a thirst for that
military fame he was destined to acquire, and the state
of the times favoured his wishes. While yet a youth,
he entered into public life at that stirring period when
the Dutch, under the guidance of his relative, William
prince of Orange, and the other members of that noble
house, were struggling for their independence, which,
after a severe conflict, they finally gained in 1648. From
that date, to a long period after, the successes of the
Dutch against their former masters, both by land and
sea, were almost uninterrupted. They attacked, with
vigour, the distant possessions of their enemies in Asia,
Africa, and America, thus acquiring, in a few years, not
only vast colonies, but immense wealth; added to a commerce which poured into their treasury " all gems in
sparkling showers." Their darling passion for commerce,
however, was not yet fully gratified, and, flushed with
success, they determined to extend their conquests in
Brazil as much as they had done in India. A rapid
sketch of their proceedings will be necessary to understand the peculiar position of count Maurice.

The Dutch, having seen the success of forming their
East India Company, determined in 1624, on establishing another for the West Indies, by which,— large
sums being soon raised in the way of shares,— they
were enabled to fit out their first expedition against
Brazil. This was commanded by admiral Willekens,
who entered the bay of Bahia, reduced its capital, and
gained immense plunder. This sudden success, however, was not lasting: the Portuguese rallied, cut off
supplies from the town, and reduced the Dutch to such
distress, that, after an obstinate defence, they were obliged to embark and return to Europe. The States,
however, had been so successful in all their other con-
flicts with Portugal, that in 1630 they determined on
making another and a more vigorous attempt to gain
that charming country. They accordingly fitted out a
fleet of no less than forty-six men of war, having a considerable body of troops, under the command of general

Waldenbourg. Three thousand men of this expedition landed at Olinda, captured that city, and thus secured the whole province of Pernambuco. By the bravery of the Dutch admirals, a powerful armament, sent out from Portugal to repel the invaders, was defeated; and in seven years the Dutch had gained possession of three other of the neighbouring provinces. The produce of these, joined to the immense riches which flowed into the coffers of the company, was such as almost surpasses belief, for it is stated * that in this period they had taken or destroyed no less than 547 vessels out of 800 sent against them by Spain and Portugal.

Such was the state of affairs when the services of Maurice were sought for by the Dutch. The West India Company, flushed with success and overflowing with wealth, resolved to attempt the conquest of the whole of Brazil, and made an offer of the command to prince Maurice, as one to whose valour and prudence every thing should be intrusted. Impatient of delay, our hero waited not for the large armament that was to accompany him, but left Holland with only 4 ships and 350 soldiers, subsequently increased, before his landing, by another 600. With only 300 men he attacked the Portuguese army, forced their camp, and entirely defeated them: with the rest of his troops he besieged the citadel of Povaceon, reduced it, and captured the garrison. The strong town of Penedu, on the Rio St. Francesco, next fell before him. Having thus gained a firm footing, he sent an expedition to capture the fort of St. George de la Mina, on the coast of Guinea, and this, although the strongest on the whole coast of western Africa, yielded to his troops. Meantime, he himself attacked and defeated the army of the chief general of Portugal, the Count de Bangola, a soldier of high reputation: he was thus put in possession both of the capital and province of Seregippa. In his next expedition, however, Maurice was not so fortunate: in an attempt to possess himself of Bahia, he was defeated by the ad-

* Grant's History of Brazil.

mirable skill and courage of the count de Bangola, so that the prince, whose force was much inferior, prudently raised the siege and returned to Olinda.

It was now, having secured such immense possessions, that Maurice sedulously occupied himself in establishing order and perfecting discipline in every department of his government; he fortified all the outposts, reviewed his troops in person, promoted those officers whose merits had been most conspicuous during the war, and gave the most liberal encouragement to such natives as would join his forces, or were inclined to live peaceably under the Dutch. By these and other wise measures he not only rendered himself highly popular with all classes, but strengthened the power of the Dutch by the best of all powers — public opinion. Meantime, although busy in establishing order in his recent conquests, he fitted out expeditions to acquire new territories. He not only again conquered the rich province of Segerippa, but also reduced the island of Loanda, on the coast of Congo ; and that of St. Thomas, which lies directly under the equinoctial line. After accomplishing these conquests, he despatched six men of war and the same number of frigates, to reduce Maragnan and the town of St. Lewis. He foresaw, that if he got possession of these points, the adjoining districts would inevitably fall. The result was another proof of his generalship ; so that, in 1641, he held seven out of the fourteen provinces of the whole empire.

But the narrow and illiberal views of the Dutch Company counteracted the wisdom and prudence of their general, and finally lost them the possession of all they had acquired through his valour and wisdom. Expecting they had secured a firm and permanent footing in Brazil, they now began to show that grasping avarice which has ever been their national character. They sent positive orders to count Maurice to adopt every means for increasing their revenue, by sending vast quantities of sugar and other produce; and they particularly enjoined him not to receive their debts in small sums, but

to enforce payment all at once.   To all these the noble
Maurice warmly and repeatedly remonstrated; but they
were deaf to his arguments.   Their sordid and narrow
minds, also, became dissatisfied with what they thought
the unnecessary profusion of his expenditure.   Opposite to
the Recief of Pernambuco, is situated a small island admirably adapted for a fortification: here, therefore, he ordered a town to be built and well fortified ; the materials being chiefly taken out of the ruins of Olinda : the
situation was admirable, and the influx of inhabitants
was so great, that he soon found it necessary to unite
this new town to the Recief by a handsome stone
bridge.*   And it thus became the centre of the Dutch
commerce.

These measures, undertaken with a view of securing
the company's capital from accidents, could not be comprehended ; and were, therefore, not relished by the
thrifty traders who composed the great bulk of the proprietors in Holland, particularly as the expense amounted
to above 40,000*l.*   But their greatest discontent arose
out of what they called the splendid palace built by count
Maurice for his own use.   This edifice, which some
writers have called " magnificent," but which, in truth,
is a very plain but substantial building (of not more
than two stories, and without any ornamental front),
is admirably situated, both as a dwelling and a fortification : it commands an extensive prospect both by sea
and land, while, on two sides, it is surrounded with gardens, in which were planted all the edible fruits that
could be procured, both native and exotic.   In front a
stone battery was erected, which rises gradually from
the waterside ; upon which are ten pieces of cannon
completely commanding the river.   The count also
erected a large villa a little way in the country, encom-
passed by gardens, adorned with fish ponds, and protected
by strong walls.   The whole was planned with so much
judgment, that the building served at once for the plea-

* This bridge we saw in the year 1807, in as perfect a state as the first
year it was built.

sure of the governor, and for the defence of the city,
which it protected on that side as by a fort.   Within
these fortifications were also laid out extensive parks and
meadows, which, by judicious management, became capable of producing every thing that was necessary for
the subsistence of the garrison ; and in the disposition
of which utility and beauty were equally combined.
Thus did Maurice expend the treasures which were the
fruits of his conquests, for the public good, which he
might easily have appropriated to his own private advantage.   But this disinterested conduct, which ought
to have insured to him the gratitude of the Hollanders,
produced a contrary effect; they finally resolved to recall him, as the only means of drawing from the colony
the largest possible revenue.   In consequence of this
order, the count sailed for Europe with a fleet of 13
large ships and about 3000 soldiers; while, in con-
formity with the economical scheme of government to
be pursued in future, only 18 companies of troops were
left to garrison all the Dutch provinces.   From the departure of this illustrious warrior and statesman must be
dated the rapid decay and final extinction of the Dutch
government in Brazil.   Count Maurice, upon relinquish-
ing the command he had exercised for eight years, left 7
provinces, 1 city, 30 large towns, 45 regular fortresses,
90 sail of vessels, 3000 regular troops, 20,000 Dutch,
60,000 negroes, and about twice that number of native
Brazilians.   But in 1655, after the expenditure of several millions of money, and the destruction of several
thousand lives, only 600 or 700 impoverished individuals finally evacuated the country, and returned to
Holland.*

It is almost inconceivable how this illustrious man,
whose life, at this period, would appear to have been
spent alternately in the camp and the council, could
find leisure even to think of science, still less to have
prosecuted it in his closet.   Yet the versatility of his
mind, and its power of abstraction, was so great, that

* See Grant's Brazil, p. 89.

such was actually the fact. He not only patronised and
assisted the labours of those whom he had engaged for
this purpose *, but actually worked himself in describing
and drawing the various new animals of Brazil, even in
the most arduous periods of his government. This remarkable trait in his character does not rest on the
equivocal or laudatory records of an historian; for there
remains, even at the present time, striking monuments
of what, with his own hand, he effected. These are
found, in the form of two folio volumes, in the Royal
Library of Berlin, containing, according to Bloch, coloured drawings of numerous animals of South America,
with a short description of each. " One part," he says,
" of these precious manuscripts is a small folio in white
parchment, containing 32 quadrupeds, 85 birds, 9 amphibia, 24 fishes, 31 insects, with several shells and
marine animals; in all 193 pages. As a whole, every
subject seems accurately designed, and partly coloured,
the colours being still vivid and beautiful. Under every
animal is its native name and dimensions. The second
part is somewhat larger; it contains 2 quadrupeds, 15
birds, 46 amphibia, 45 fishes, 46 insects, and many pages
with plants; in all 114†;" the whole of which were
executed by the hand of Maurice. Marcgrave, who was
more immediately in his service, could not have laboured
with greater assiduity, although he explored a wider field.
He visited the countries from between the banks of the
Rio Grande and Pernambuco, collecting many observations
on geography and astronomy, as well as in natural history.
He left Brazil the same year as his patron, proceeded to
the Coast of Guinea, and there found an early tomb.
Under the direction of Maurice, those of his papers
which bore on astronomy and geography were published
by professor Golius, and those on natural history, along
with Pison's, were put into the hands of John de Lact,
and printed at the Elzevir press, at the expense of his
princely patron. Marcgrave had, himself, so far accom-

---

* Marcegrave and Cralitz.
† *Ichthyologie*, par M. E. Bloch, 1788, 6 part in præfat.

plished the task of authorship before his death, as to
dedicate the work to him from whose munificence it
originated.

Although recalled by the States, the services he had
rendered were too important to pass unrewarded ; he was
accordingly appointed governor of Wesel, and general-in-
chief of the Dutch cavalry. The elector of Branden-
burg soon after named him Grand Master of the Teutonic
Order, and made him Governor of the Duchy of Cleves,
where, honoured and respected, he seems to have spent
the remainder of his days. His love of nature and the
fine arts, long survived the thirst for conquest and
martial fame. Retired from such empty pageants, he
greatly adorned the town, and founded a magnificent
garden. Voltaire, writing a hundred years after from
Cleves, remarked:—" It is the loveliest spot in nature,
and art has done much to improve it. I shall only
add, that a certain prince Maurice of Nassau, who was
governor of this beautiful solitude, was the principal
agent in accomplishing all these marvels." *   In this
delightful retreat, and at the advanced age of seventy-
five, he expired on the 20th of December, 1679. †

MAWE, JOHN.—*Conchology.*

A diamond merchant and commercial dealer in shells,
&c., settled in London. He was the first Englishman
who visited the gold and diamond districts of Brazil ;
an account of which he published on his return : he is
likewise the author of several useful little tracts on the
elements of his own profession.

1. *The Shell Collector's Pilot ;* pointing out where
the best Shells are found in all Parts of the World.
1 vol. 18mo.

2. *The Linnæan System of Conchology,* with 36

* Voyage à Berlin.
† There are three portraits of Count Maurice ; the best is by T. Matham,
in folio ; another, rather larger, by P. Soutman ; and the third, 12mo. size,
without the name of the engraver, is inscribed, *J. Mauritius Comes Nas-
sovius, Generalis in Brasilia.*

Plates drawn on Stone by E. A. Crouch. Lond. 1823.
1 vol. 8vo. The figures, in general, are remarkably
good, and will always render this work valuable to
students, notwithstanding its obsolete nomenclature.

MAXIMILIAN, PRINCE.— *General Zoology.*

Prince of Weid Neuwied; an enterprizing traveller
and zoologist, whose love for science led him to explore
the interior of Brazil. He has published, according to
Cuvier, the following works:

1. *Voyage to Brazil,* 2 vols. 4to. with an Atlas.
Frankfort, 1820, 1821.

2. *Natural History of Brazil.* 2 vols. 8vo. Weimar,
1826. Also a few numbers of coloured plates in
folio. A translation, in quarto, appeared in London,
of the first volume of his travels.

MECKEL, J. F.— *Comparative Anatomy.*

Professor at Halle.

Materials for advancing Comparative Anatomy (in
German). Leipzig, 1808. 8vo.

MEIGEN, J. W. — *Entomology.*

An acute and indefatigable entomologist, now living,
who has devoted his researches exclusively to the dipterous insects ; upon which order he is the first authority.

1. *Nouvelle Classification des Mouches* à deux Ailes.
Paris, 1800. 8vo. This, which may be considered
as the Prodromus to Meigen's great work, was published by M. Baumhauer. In 1803, Illiger likewise
gave, in his Magazine of Insects, a synopsis of Meigen's system.

2. *Klassificazion* und Beshreibung der Europæ-
ischen Zweiflügligen Insecten. Bruns. 1804. 2 parts
in 1 vol. 4to. fig.

3. *Systematische* beschreibung der bekannten Eu-
rop. Zweiflügeligen Insecten.   Aachen, 1818—37.
7 vols. 8vo.

MERIAN, M. S.—*Entomology.*

Maria Sibilla Merian, deserving of honourable mention among those who have advanced the interests of
natural history, in consequence of the admirable manner
in which she delineated insects and other natural objects, was born at Frankfort, in the year 1747.   She
employed her early years in portrait and miniature
painting, under Abraham Mignon ; but, while prosecuting these departments of the art, she neglected no
means of improving her skill in figuring plants and
insects.   She was early married to a painter of Nuremberg, named John Andrew Graf ; from whom, however,
she soon separated.   Her first work, illustrative of the
metamorphoses of insects, appeared in 1679, and is
entitled " Erucarum Ortus, Alimenta, et Paradoxa Metamorphosis."   This work has been translated into
German and French ; and although the drawings are by
no means equal to those she subsequently produced, it
is curious, both by representing the various stages of
numerous insects, and from the circumstance of some
of the engravings being executed by her own hand.
Madame Merian was an enthusiast in religion, and became for a time a convert to the opinions of Anna
Maria Schurman.   In the year 1699, she visited Surinam, for the purpose of delineating the insects of
Tropical America ; the splendour of which, as she saw
them in the cabinets of Holland, had excited her admiration. She returned to Europe in 1701 ; and the beautiful
drawings she brought with her were published in 1705,
under the title of " Metamorphosis Insectorum Surinamensium ;" the text by Caspar Commelin.   This
magnificent work greatly surpassed any previously devoted to such subjects ; and although it contains not a
few serious errors, it may even now be consulted with
advantage by the entomologist.   The grouping of the

various objects brought together on the plates displays
great artistical skill.   This accomplished and enthusiastic lady died on the 13th of January, 1717. — J. D.

1. *Metamorphosis Insectorum Surinamensium.*
Ant. 1705—1709. folio, 60 plates.   The text is in
French and Dutch; the plants are described by
Caspar Commelin.   Some copies are most splendidly
coloured by herself.   There is another edition, printed
at the same time, with the text in Latin.   These,
the original editions, seem to be unknown to Cuvier,
who merely alludes to " two posthumous works."

2. *De Generatione* et Metamorphosibus Insectorum Surinamensium. Hagæ Com. 1726. folio. The
number of plates is 72 ; and the text is in Latin
and French.

3. *The Papilionaceous Insects of Europe,* with
their Transformations, and the Plants upon which they
feed (in German). Nurnb. 1679—1683. 2 parts in
1 vol. 100 plates.   A translation of this work is
thus mentioned by Cuvier: — " Histoire des Insectes
d'Europe, trad. fr. par Mairet.   Amst. 1730. 1 vol.
folio."

MERREM, B. — *Ornithology and Erpetology.*

Born at Bremen.   Professor of Natural History at
Marpurg.   One of the early reformers in the arrangement of the reptiles.

1. *Avium rariorum* et minus cognitarum Icones
et Descrip.   Leipsig, 1786. 4 numbers in 4to.

2. *Materials* for a Natural History of Reptiles
(in German). Duisbourg, 1790. 2 numbers in 4to.

MIKAN, J. C. — *Entomology.*

Monographia Bombyliorum Bohemiæ.   Prague,
1796. 8vo. pl.

T

MILLER, J. S. — *Malacology.*

Late curator of the Bristol Museum ; an accurate and industrious observer of nature.

> *Natural History of the Crinoidea,* or Lily-shaped Animals, with Observations on the Genera Asterias, Euryale, Comatula, and Marsupites. 1 vol. 4to. 40 coloured plates. Bristol, 1821.

He is also the author of a memoir on the belemnites, published in the Transactions of the Geological Society of London, vol. ii. 2d series, part i.

MITCHELL, DR. S. L. — *Ichthyology.*

A distinguished American ichthyologist, who has written much on the fishes of North America ; but his papers are chiefly inserted in Transactions, and his works not well known in this country. The following is noticed in the Linnæan Society's Catalogue.

> *On the Fishes of New York.* 12mo. New York, 1814.

MOEHRING, P. H. G. — *Ornithology.*

> *Avium Genera.* Aurich. 1752. 8vo.

MONTAGU, GEORGE. — *General Zoology.*

A gentleman of fortune, and a colonel in the army, who was zealously devoted to natural history, and has left some valuable descriptions of animals discovered or observed by him on the western coast of England. He was an accurate observer of nature, and one of the most eminent of our native zoologists.

> 1. *Ornithological Dictionary,* or Alphabetical Synopsis of British Birds. 2 vols. 8vo. London, 1802. A second edition was published by Rennie, in 1831,

overlaid with much commonplace matter, and with little improvement on the original: the woodcuts to this edition are very poor performances.

2. *Supplement* to the Ornithological Dictionary, or Synopsis of British Birds. Exeter, 1813. 8vo. 1 vol. plates 23. This valuable work contains more information respecting our native birds than is to be found in any others yet published, excepting, perhaps, Mr. Selby's.

3. *Testacea Britannica,* or an Account of all the Shells hitherto discovered in Britain. 2 vols. 4to. with plates (drawn and engraved by his friend Mrs. Dorville). Romsey, 1803. To this a Supplement was afterwards added, in thin quarto. London, 1808.

Several valuable papers on birds, fishes, and *Mollusca,* are likewise inserted in the Linnæan and Wernerian Transactions, mostly illustrated with very good figures by the same pencil.

MONTFORT, DENIS DE. — *Conchology.*

Of this eccentric, but by no means inaccurate, writer, Cuvier remarks, — " This singular man styled himself an ancient naturalist to the king of Holland." He seems to have committed some offence, for which (as Dr. Leach once told us) he was condemned to the galleys. Cuvier says he perished through want, in the streets of Paris, in 1820 or 1821. From such a shocking fate, whatever his crimes might have been, he surely ought to have been saved.

Conchyliogie Systématique et Classification Méthodique des Coquilles. Paris, 1808—10. 2 vols. 8vo. The figures are, in general, ill drawn and worse engraved. By some strange forgetfulness (for we cannot suppose it ignorance), Cuvier says that these figures " are as exact as can be done by that species of engraving!" Yet the author is by no means destitute of talents ; and, notwithstanding its barbarous

T 2

Latin names, the book contains many new and accurate observations.

MOQUIN, DR. A. TANDON. — *Entomology.*

A physician, and one of the Professors at Marseilles.

Monographie de la Famille des Hirundinées. Mont. 1826. 4to.

MORREN, C. F. A. — *Entomology.*

De Lumbrici Terrestris Historia Naturali, necnon Anatomia. Brussels, 1829. 1 vol. 4to.

MÜLLER, O. F. — *Zoology and Botany.*

One of the most accurate and laborious naturalists of the eighteenth century; but of whose biography we know nothing more than that he was born in 1730 — was created a counsellor of state — and died in 1784.

1. *Zoologiæ Danicæ* Prodromus. Hafniæ, 1776. 8vo.

2. *Hydrachnæ.* Lipsiæ, 1781. 4to. col. pl.

3. *Entomostraca* seu Insecta Testacea. Lips. et Hafniæ, 1785. 1 vol. 4to. fig.

4. *Animalcula* Infusoria. 1 vol. 4to.

5. *Von Würmern der Süssen* und Salzigen Wassers, i. e. On Freshwater and Marine Worms. Copenh. 1 vol. 4to.

6. *Vermium* Terrestrium et Fluviatilium Historia. 2 vols. 4to.

7. *Zoologia Danica.* Copenh. 1788. Folio, with coloured plates. The first parts are by Müller; it has been since continued by various hands.

NATTERER, ——. — *Ornithology.*

An acute and most zealous ornithologist, who was engaged by the government to explore Brazil, and transmit his collections to the Vienna Museum. He left Europe in 1816 or 1817, and, after travelling that vast empire, in almost all directions, for nearly sixteen years, returned to his native country, by the way of England, in 1835. He is now at Vienna, and the scientific world is anxiously expecting the publication of the results of his long and arduous travels. Previous to his expedition, he had paid great attention to the birds of Europe, and had discovered several new species, since described in Temminck's Manuel. His Brazilian collections must be immense, as he assured me he had found more than 1000 species in that region!

NAUMAN, J. A. AND J. F., FATHER AND SON. — *Ornithology.*

Conjointly the authors of an " excellent work," says Cuvier, on the *Natural History of the Birds of Germany,* " the plates of which, though small, are perfect." A second edition, in 8vo., was commenced at Leipsic in 1820.

NEIREMBERG, J. E. — *General Zoology.*

Professor in the Jesuits' College at Madrid.
Historia Naturalis maxime peregrina, Lib. 16. distincta. Anvers, 1633. folio.

NICHOLSON. — *Traveller. Natural History in general.*

An Irish Catholic, who lived, as missionary, some years in St. Domingo. His work has little pretensions to science, but contains many interesting observations: it is now scarce.
Essai sur l'Histoire Naturelle de St. Domingue. Paris, 1776. With a few plates.

NILSON, S. V.— *Ornithology.*

Ornithologia Suecica. Copenhagen, 1817—21. 2 vols. 8vo. A work of some repute. The author was curator of the museum at Lund.

NOZEMAN, CORNELIUS.— *Ornithology.*

Nederlandsche Vogolen, &c., or the Birds of the Low Countries, with their Nests and Eggs, described (in Dutch) by C. Nozeman, and continued after his death by Martin Houttuyn ; the Plates designed, engraved, and coloured from Nature by Christ. Sepp. Amst. 1770—1829. 2 vols. folio, complete, with 250 plates. This magnificent work is not often seen. Cuvier observes that it is remarkable for the elegance of its figures. The selling price is about 30*l.*

OCHSENHEIMER, F.— *Entomology.*

One of the best entomologists who have written upon lepidopterous insects. His work on the *Lepidoptera* of Europe, in which a great many new genera are proposed, is in German. The four first volumes are by himself, the others by Treitschke : there are no plates. An English translation, with notes, highly useful to British entomologists, was begun by Mr. Children, late of the British Museum ; but we know not whether it has been finished.

Die Schmetterling von Europa. 8 vols. Leipzig. 8vo. 1807—1830.

OKEN.— *General Zoology.*

A celebrated naturalist of Germany, and chief editor of the *Isis,* the best journal in Europe on natural history. Yet so little are his writings known in this country, that we have never been able to meet with them, either in public or private libraries, or even in booksellers' catalogues. One reason, perhaps, for this, originates in their being written entirely in German. We therefore quote the following from Cuvier: —

1. *Philosophy of Nature*, 3 vols. 8vo. Jena, 1809.

2. *A Treatise on Natural History ;* of which the Zoology forms the Third Part. 2 vols. 8vo. with Atlas. Jena, 1809.

3. *A Natural History for Schools*, 1 vol. Jena, 1821.

4. *Esquisse de Systéme* d'Anatomie, de Physiologie, et d'Histoire Naturelle. Paris, 1821. 8vo.

OLINA, GIO. PIETRO. — *Ornithology.*

Uccelliera overo Discorso della Natura e Proprietà di diversi Uccelli e in particolare di quelli che cantono, &c. Opera di Gio. Pietro Olina. Rome, 1622. 1684. folio.

OLIVI, ABBATO GUISEPPE. — *Malacology, &c.*

Zoologia Adriatica. Bassano, 1792. 4to. pl. 9. pp. 334. The figures are very good.

OLIVIER, ANTOINE GUILLAUME. — *General Natural History.*

Although of varied and extensive acquirements, Olivier is now principally remembered as an entomologist. He was born at Arcs, near Frejus, on the 19th of January, 1756; and went to Montpellier, where he obtained his medical degree at the early age of seventeen. The very slight acquaintance with certain branches of natural history which the ordinary medical curriculum of study requires, frequently suffices to inspire a taste for it; and hence it is that so many medical men become naturalists. It was thus that Olivier became attached to the pursuit; and his predilections were strengthened

by intercourse with Broussonet, who was then residing at
Montpellier.

On leaving that place, Olivier returned to his native
district, and began practice as a physician; but his
success not being encouraging, he spent most of his time
in studying plants and insects. Meanwhile, Broussonet
was interesting himself on his behalf in the capital; and
when Berthier de Savigny, intendant of Paris, formed a
scheme for including an account of the natural productions of the vicinity of that city in a statistical and
economical work, Olivier was chosen, on his friend's recommendation, to undertake the task. This he executed
in such a manner as to show that he was capable of
greater things; and a favourable opportunity soon offered. Gigot d'Orcy, a wealthy financier, who cultivated
entomology, projected a general work on that subject;
and he employed Olivier to travel through Holland,
England, and other places, to collect specimens and make
drawings. We are not aware that the original scheme
was carried into effect; but the knowledge thus obtained
by Olivier, enabled him to draw up those elaborate and
valuable treatises contained in the *Encyclopédie Méthodique*, and to lay the foundations of his great work on
the *Coleoptera*. It having been determined, by one of
the parties which held temporary authority during the
revolution, that a mission should be sent to Persia to
establish commercial relations with that country, Olivier
and Bruguière were commissioned to accompany it in the
capacity of naturalists. On this expedition he was engaged
six years, during which he suffered much inconvenience,
in consequence of the minister, Roland, the projector of
the mission, having been driven from power, and his
successors disregarding the object he had in view. He
returned to France in December, 1798, bringing with
him large collections in natural history. He then employed himself in drawing up an account of his journey,
and in continuing his work on insects. Being possessed
of some fortune, he could spend his time according to
his inclinations; and we do not find that he occupied

any situation of consequence, except that of Professor of
Zoology in the Veterinary School of Alfort. He was
elected a member of the Institute on the 26th of January, 1800. In the latter part of his life, his health, which
had been very robust, gave way, and he travelled through
different parts of Europe to restore it; but he was
found dead in his bed at Lyons, on the 1st of October,
1814, his disorder proving to be aneurism of the aorta,
the existence of which had not been suspected by his
physicians.

Olivier's subordinate works, which relate to various
subjects in botany, entomology, and agriculture, are to
be found in the various periodicals of his day; particularly the *Mémoires de l'Institut, Mémoires de la Société
d'Agriculture, Journal d'Hist. Nat., Feuille du Cul-
tivateur*. The great work on *Coleoptera* is in six quarto
volumes (1789—1808), with 363 plates. It has long
been a standard work of reference for that extensive
order of insects. — J. D.

1. *Entomologie*, ou Histoire Naturelle des Insectes.
Paris, 1789—1808. 5 vols. 4to. Complete copies of
this laborious work are so very rare, that authors
differ as to the number of its volumes. Cuvier cites
only 5; Kirby and Spence, 8. All the copies we know
of are imperfect. The plates are numerous, well
filled, and sufficiently well executed and coloured.
The descriptions, so far as they go, are remarkably
accurate. The work does not extend beyond *Coleoptera ;* but it is the most complete account of those
insects that has, or perhaps ever will be, published.

2. *Voyage* dans l'Empire Ottoman, l'Egypt, et la
Perse. Paris, 1807. 3 vols. 4to. The animals observed by the author are described with great accuracy.

OPPEL, MICHAEL. — *Erpetoloy*.

A very able and excellent erpetologist, native of Bavaria, whose writings are still of much value.

1. *Sur la Classification* des Reptiles. Two essays in the Annales du Muséum.

2. *The Orders, Families, and Genera* of Reptiles (in German). Munich, 1811. 1 vol. 4to. He wrote other papers in scientific Transactions; and commenced a great work on reptiles, which was not continued.

OSBECK, PETER. — *Traveller. Zoology and Botany.*

One of the celebrated pupils of Linnæus, who went to China as chaplain in a Swedish vessel in 1750. His narrative was originally printed in the Swedish language, in 8vo. at Stockholm, in 1757: but there are two translations: one into German, by Rostock, 8vo.; the other by Forster.

A Voyage to China and the East Indies; to which is added, a Faunula and Flora Sinensis. 2 vols. 8vo. with plates. London, 1771.

PALLAS, PIERRE SIMON. — *General Zoology.*

Peter Simon Pallas enjoyed in early life the parental superintendence of those who were alike willing and qualified to fulfil the important task of education. His father, professor Pallas, who had been bred to the medical profession, attained the rank of surgeon-major in the army; and, on settling in Berlin, was appointed professor of surgery and chief surgeon to the public hospital. It was in this capital, on the 22d of September, 1741, that Pallas was born. His early education was conducted under his father's roof, with the assistance of private tutors, and was distinguished by extraordinary attainments in languages; so that, when almost a child, he could not only speak, but accurately write, in Latin, and French, and English, as well as in his native tongue. His father destining him for his own profession, in his fifteenth year he commenced its study, in the Berlin University, under his father's auspices, and with the assistance of such able coadjutors as Meckel and Sprengel.

Thus was a foundation laid in an extensive curriculum ; and so assiduously did Pallas avail himself of his peculiar advantages, that, at the age of seventeen, he gave lectures on anatomy. He did not, however, confine his attention to these professional studies, but found leisure to attend to several departments of zoology, more especially to entomology.

After thus assiduously availing himself, for between three and four years, of the opportunities which were presented in the Prussian capital, at his father's suggestion, Pallas sought the advantages of foreign travel; and in 1758 set out upon a protracted visit to the most celebrated universities and medical schools. He first repaired to Halle, where his attention was engaged with mineralogy, mathematics, and physics : thence he went to Göttingen, where he spent much of his time upon comparative anatomy, the action of morbid poisons, and parasitical animals : in 1760 he resorted to Leyden ; and here, at the age of nineteen, took his degree of Doctor of Medicine. Holland was at this time luxuriating in her commercial splendour ; and science, as a consequence, was daily deriving new objects of wonder from every quarter of the globe. Her museums were unrivalled ; and it was at this time, and under these circumstances, that a bias was given to Pallas's genius, and that natural history, in its widest sense, became the object of his peculiar predilection. In the year 1761, the young physician passed over to London, with the ostensible object of attending its hospitals ; but, to say the least of it, to this he conjoined the pursuit of natural history, —and that, not merely in the study, but in the wide field of nature, making various excursions to the sea-coast, and examining the rich productions of the ocean.

Professor Pallas now wishing his son to return and practise the healing art in the bosom of his family, he arrived in Berlin in June, 1762. But the fair field of Nature's works had excited his ardour, and his thirst for their cultivation was unquenchable. After repeated entreaties, he obtained his father's consent to remove and settle at the Hague, where the stadtholder's splendid collection of natural history was being arranged, — and than which, a better school could not possibly have been selected. Here he remained three years; finished the education of his early prime, and planted the seeds of knowledge which soon yielded so rich a harvest: here his value began to be felt, and his rising powers cherished, — in his own words, " summa humanitate a curiosis et scientiæ patronis excerptus fui." In the year 1764, at the age of twenty-three, he was elected F.R.S. and member of the society called *des Curieux de la Nature.*

In the year 1766, the young naturalist published two works, both remarkable for the observations they communicated, and the large views they displayed. These were, the *Elenchus Zoophytorum* and the *Miscellanea Zoologica;* the former characterised by Haller as *princeps hac classe opus;* and though in it he displayed his ignorance of the more successful labours of our distinguished countryman Ellis, and even treated him disparagingly, yet he soon honourably atoned for his blunder — *Ellisium subtilitate observationum omnes supereminentem.* The *Miscellanea* was scarcely less remarkable; containing interesting views, more especially regarding the *Mollusca,* all but opening up a new era in their history, and contributing a storehouse upon which the author drew for several of his subsequent works.

At this time Pallas presented a plan of a scientific expedition to the foreign Dutch settlements to the prince of Orange, and offered to superintend the undertaking. The only barrier was the want of his father's consent ; and he was recalled, reluctantly, to Berlin. Here he began his well known *Spicilegia Zoologica ;* and in a few months four numbers were published.

In the year 1767, Pallas was invited by the Russian monarch to become Professor of Natural History in the Petersburg Academy of Science ; this he accepted, and although his inclinations were again opposed by his father, he reached the Russian capital in August of that year. Here, however, his stay was short ; the empress Catherine having planned that a scientific expedition should proceed into Siberia, to which five principal naturalists were attached,—Lépécher and the younger Gmelin being two. Pallas drew up the general instructions for the others, and selected the country beyond the Volga for his own investigations. Before all, however, was ready, he found time to publish six additional numbers of his *Spicilegia,* and a memoir on the fossils of Siberia, which created a great sensation throughout Europe. We cannot, of course, follow him through the laborious journeys to which many of his associates fell a sacrifice, and in which he displayed an energy which could not be exceeded. They toiled for the long period of six years, during which he penetrated as far as the frontiers of China, and traversed from the Caspian to a high latitude. According to a preconcerted plan, he sent every year, ready for press an account of his travels to Petersburgh, which were immediately published, and translated into French,—forming journals, which from competent judges have received the highest praise. After his return to Petersburgh, in 1774, he began to digest his materials, and many were the works which proceeded from his pen ; some on the collateral branches of science, but the greater number on natural history, from man down to insects : he began, under the auspices of the empress, a *Flora Russica,* and contemplated even a *Fauna Russica.* On resuming his post, honours thickly showered upon him, of which we can only name that he was appointed a member of the Board of Mines, with a salary of 200*l.* a year : the empress purchased his collection for 20,000 rubles, raised him to the order of Vlodimir, and made him an imperial counsellor.

After thus spending twenty years in the Russian metropolis, in the prosecution of his honourable calling, the conquest of the Crimea presented the occasion of fresh travel. He renewed this delightful recreation in the year 1793, and spent nearly two years in the southern

provinces of Russia. Enchanted with ancient Taurida,
Pallas now came to the determination to quit Petersburgh, and return to this celebrated retreat ; and here
the empress exhibited her wonted regard, — presenting
him with large domains, and a pecuniary gratuity for
building a mansion. This accomplished, with his habitual energy, he planned agricultural improvements for
the country, and prepared an account of his recent travels, which appeared in German, French, and English;
one of the very few of his works which have been
honoured by receiving an English dress. Soon, however,
in this choice retreat, Pallas discovered that unforeseen
sources of annoyance opened up which could not be removed. The country being still in a disturbed state, the
people were so rude as greatly to thwart all his plans of
improvement ; he was likewise harassed with legal ques-
tions respecting his property ; and hence we find him,
in the year 1801, complaining of the many vexations to
which he was exposed. These he was urged to leave
far behind ; but it was not till 1810 that he disposed
of his property in the Crimea, and returned to his native land. At this time, he was still in the enjoyment
of a large share of health; and in the renewal of old associations, in the unbounded respect with which he was
every where entertained, in the delights of scientific
information, and in the domestic sweets of a brother's
and a daughter's society, he experienced a large measure of bliss. He now planned an extensive scientific
tour into France and Italy ; but this scheme was cut
short by an unexpected attack of alarming disease,
which snapt the thread of life at the venerable age of
seventy.

He.was twice married, and left an only daughter, who
became baroness Wimpfen. Part of his valuable collection he bequeathed to the university of Berlin. — J. D

1. *Miscellanea Zoologica.* La Haye. 4to.

2. *Spicilegia Zoologica.* Berl. 1767—1780. 14
Nos. in 4to.

3. *Novæ Species* Quadrupedum e Glirium Ordine,

cum Illustrationibus variis complurium ex hoc Ordine Animalium. Erlang. 1778. 4to. pp. 388. pl. 39.

4. *Travels* through several Provinces of the Russian Empire (translated both into English and French).

5. *Neue Nordische Beytraege*, &c. (New Materials on the Geography, &c. of the North). Petersb. et Leipzig, 1781—1796. 7 vols. 8vo.

PANZER, G. W. F.—*Entomology.*

An indefatigable entomologist and laborious writer. He was a physician at Nuremberg, and born in 1753.

1. *Faunæ Insectorum Germanicæ* Initia (Deutschlands Insectens). In 109 small cases or numbers in 12mo., each containing 24 col. plates. Nuremb. 1796. This is one of the most useful illustrative works in the science, both from the multitude and exactness of the figures, and the accuracy of the nomenclature.

2. *Entomologischer* versuch über die Jürineschen Gattungen der Linneischen Hymenopteras. Nurnberg, 1806. 1 vol. 12mo.

3. *Index Entomologicus.* Pars prima, Eleutherata. Nurembergæ, 1813. 1 vol. 12mo. pp. 216.

4. *Kritische* Revision der Insectenfaune Deutschlands. Nurnberg, 1805—6. 2 vols. 8vo. 2 col. pl.

5. *Iconum Insectorum* circa Ratisbonam indigenorum Enumeratio Systematica. Erlang. 1804. 1 vol. 4to. pp. 260.

PARNELL, DR.—*Ichthyology.*

A physician, established at Edinburgh. One of the most eminent ichthyologists of Britain.

The Natural History of the Fishes of the River Forth, in Scotland; published in the Wernerian Transactions, and which gained the annual prize established by that learned body. Dr. Parnell, by this invaluable essay, has placed himself at once in

the foremost ranks of living ichthyologists. A few
copies have been printed separately, for private distribution. Dr. Parnell is now in the West Indies;
where he will, no doubt, find leisure to prosecute his
favourite science, and from his hands we cannot but
expect some most valuable results.

PARRA, DON ANTONIO. — *Zoology and Ichthyology.*

One of the few really good naturalists of which Spain
can boast. He visited, or resided at, Cuba; and his
work, written in Spanish, is much sought after, on
account of the figures and descriptions it contains of
fishes no where else mentioned. We have never been
able to consult or see it, as it has become exceedingly
rare.

   *Description of various Portions of Natural His-
tory,* and chiefly of Marine Productions. In Spanish.
1 vol. 4to. Havanna, 1784.

PAYKULL, G. — *Entomology.*

Councillor· of state, and Member of the Royal Academy at Stockholm. He has investigated the insects of
his native country with great acuteness.

   1. *Fauna Suecica;* Insecta. Upsaliæ, 1800. 3 vols.
8vo.
   2. *Monographia* Histeroidum. Upsaliæ, 1811.
1 vol. 8vo. with figures of all the species.

PENNANT, THOMAS. — *General Zoology.*

Thomas Pennant, whose name is so familiar to the
British zoologist, was one of the first who successfully
investigated the history of our native animals. He was
descended from an ancient and honourable family of
Wales, some of whom traced their descent from the
great Madoc; while others date it, with a greater show

of probability, from Richard Plantagenet, duke of
York. His immediate predecessors belonged to a branch
of the family of Hugh Pennant, of Bychon. He himself was born at Downing, in June, 1726. According
to his own account, he was first imbued with a love for
natural history at twelve years of age, by having a copy
of that fine old volume, Willughby's *Ornithology*, put
into his hands. When he reached the years of man-
hood, he spent some time in making tours through various parts of Britain and Ireland, directing his attention chiefly to the study of geology and mineralogy.
In 1757, on the recommendation of Linnæus, he was
elected a member of the Royal Society of Upsal, which
he always regarded as the greatest of his literary honours. He commenced his " British Zoology " in 1761;
originally on a large scale, but smaller editions were
subsequently published. A lengthened journey on the
Continent gave Pennant an opportunity of becoming
acquainted with Buffon, Haller, Pallas, and many other
naturalists of that day, whose correspondence he enjoyed
and cultivated in after years. On returning home, he
commenced an " Indian Zoology;" which, however, he
abandoned, after fifteen of the plates had been engraved,
partly at the expense of sir Joseph Banks: the reasons
for this step are not known ; but it seems these plates
were afterwards published in Germany by Reinhold
Forster. In June, 1769, he made an excursion to
Scotland, with a view to explore its natural productions.
His tour in that country, with which he was much delighted, was published, as well as that of a subsequent
and more extended expedition, undertaken in the summer of 1772. He then traversed various parts of England and Wales, — investigating the history and antiquities of every place he visited, more, perhaps, than its
natural productions. " Arctic Zoology " was the next
work of importance he undertook, and this occupied
much of his attention. " A History of London," and
" Outlines of the Globe," also emanated from his pro-
lific pen ; the latter, conceived on an extensive scale,

U

never went beyond the fourth volume, and two of these were posthumous. The equanimity of his disposition, and extreme activity of mind, were not interrupted by an accident which befell him in his latter years, and he continued his literary and scientific labours almost to the last. His death took place on the 16th of December, 1798, at the age of seventy-two. — J. D.

The fame of Pennant must not be measured simply by his talents as a naturalist; in that he merely kept pace with the then state of science, and did nothing to improve its philosophy. But he united in himself what few men, and still fewer naturalists, possess, — namely, an enlarged and elegant mind, richly stored with classic lore, and with extensive and varied reading: hence he possessed that happy facility of interesting the reader on matters which, in other hands, would have been dull and technical. Pennant, in short, was the elegant scholar and the refined gentleman; and as such, his place has never yet been filled in the annals of our science up to this day. Had his powers been concentrated on zoology alone, we doubt not, he would have equalled the most eminent of his age; but, by being diffused over so many subjects, they became diluted. His character, in fact, is one of rare occurrence — uniting the greatest application with the most disinterested love of literature; for he held a station in society, which rendered him above the daily duties of a professional authorship. Whatever he touched, he beautified, either by the elegance of his diction, the historic illustrations he introduced, or the popular charm he gave to things well known before. — W. S.

1. *The Genera of Birds.* London, 1781. Thin 4to. pp. 68. and xxv. with plates.

2. *Indian Zoology.* London, 1790. 4to. 16 plates.

3. *Arctic Zoology.* 3 vols. 4to. Second edition, 23 plates. London, 1792.

4. *British Zoology.* 4 vols. 8vo. London, 1768 —1777; of this there are several editions.

5. *History of Quadrupeds.* 2 vols. 4to. 3d edit.
109 plates. London, 1793.
6. *Synopsis* of Quadrupeds. 8vo. Chester, 1771.

PERCHERON, A. — *Entomology.*

An able entomologist of France. Born in Paris, January 1797, and author of the most valuable bibliographic work on his favourite science that has yet appeared; besides the following, he has written several papers in Guérin's Zoological Magazine.

1. *Bibliographie Entomologique,* comprenent l'Indication de Noms d'Auteurs, &c. Paris and London, 1837. 2 vols. 8vo.

2. *Monographie des Cétoines* et Genres voisins, in conjunction with M. Gory. Paris, 1833. in 8vo. with coloured plates, complete in 15 numbers.

PERNETTI, DOM.

An ecclesiastic of the order of St. Benedict, who accompanied Bougainville to the Molucca Islands. On his return, he was appointed librarian to Frederick II.

Voyage aux Isles Malouines. Paris, 1770. 2 vols. 8vo. The occasional notices on the natural history of the country, and the figures of several rare animals, render this a work of frequent reference.

PÉRON, FRANCOIS. — *Zoologist and Traveller.*

A celebrated zoologist and traveller, who accompanied one of the French circumnavigators, and has enriched science with many valuable essays and discoveries. He was the companion of Le Sueur; and they conjointly meditated a splendid work on the marine animals found on their voyage, which unfortunately was never carried into effect. Born in 1775; died 1810.

Voyage de Découvertes aux Terres Australes, fait par Ordre du Gouvernement, par les Corvettes Le Géographe, Le Naturaliste, et Le Casuarina, pendant les

Années 1800—4, rédigé par M. Péron, et continué par M. Louis de Freycinet. Paris, 1807. 4 vols. 8vo. and one 4to. atlas of plates. Another edition was published in 1834, with the addition of 25 plates; thus making the total number 68. Péron contributed more than any one on this voyage to the collections of the French Museum.

PETAGNA, VINCENZO. — *Entomology.*

Professor at Naples; a learned and ingenious ecclesiastic. His writings are chiefly botanical; yet he studied the insects of his own country, and has described many new species.

1. *Specimen Insectorum* Ulterioris Calabriæ. Franco. 1787. A thin quarto pamphlet, with one plate.

2. *Institutiones Entomologicæ.* 2 vols. 8vo. Napoli, 1792.

PETIVER, JAMES. — *General Natural History.**

A wealthy apothecary of London, cotemporary of sir Hans Sloane, and, like him, an ardent collector. Sir Hans purchased his museum for 4000*l.* His works are curious, and his figures would be oftener quoted were the plates not so scarce. He died in 1715. His chief works are,

1. *Museum Petiveranium.* Centuria 1—10. folio and 4to. London, 1695—1703.

2. *Gazophylacium* Naturæ et Artis. 6 decades, with 100 plates, folio. London, 1702—1711.

3. *Opera* Historiam Naturalem Spectantia. 2 vols. folio. London, 1764—1773 (a posthumous work?), with 306 plates.

PFEIFFER. — *Conchology.*

A German malacologist, whose work on the freshwater shells evinces great exactness and talent. The

* See Preliminary Discourse, p. 32.

plates are coloured, and the figures of the shells remarkably exact.

Systematische Anordung, &c. or an Account of
the Land and Fresh-water Shells of Germany. 1 vol.
4to.   Cassell, 1821, with 24 coloured plates.

PHILIPPI, R. A. — *Conchology.*

Enumeratio Molluscorum Siciliæ.   1 vol. 4to.
Bresl. 1836.

PLANCHUS, J., OR BIANCHI.— *Conchology.*

Physician at Rimini. Born in 1693; died in 1775.
De Conchis minus notis. Venezia, 1739. 4to. pl.
A second edition, greatly enlarged, was published at
Rome in 1760.

PLUMIER, C. — *Botany and Zoology.*

Charles Plumier, one of the most renowned botanists
of his day, and one of the most indefatigable travelling
naturalists that ever lived, was born at Marseilles, in
the year 1646.   After greatly distinguishing himself in
his early studies, particularly in physics and mathematics, he joined the monastic order *des Mimes,* at the
early age of sixteen.   His attachment to botany brought
him into terms of intimacy with his cotemporaries Boccone
and Tournefort, with the latter of whom he became
very intimate, and was his companion in many of his
herborising excursions. Plumier's zeal and attainments
were not long in bringing him into notice ; and when
the king of France resolved to send a naturalist to his
possessions in the Antilles, in order to examine and
collect their productions, Plumier was recommended as
the person most competent to undertake that task. The
expedition set sail in 1689; but Plumier disagreed with
some of the authorities, and separated from it at the
end of eighteen months.   On his return, he published
his first voyage, which met with a very favourable

reception. Induced partly by the discoveries which
Plumier had made, and partly by the report of the many
interesting objects arriving from the New World in
other parts of Europe, particularly in Britain, the French
monarch again made arrangements to have him sent to
America. This voyage was succeeded by a third, in
which he visited the principal of the West Indian
islands,—particularly Guadaloupe, Martinique, and St.
Domingo,—and returned laden with natural curiosities,
as well as an immense number of manuscripts and
drawings. Great curiosity then prevailing among medical men to become acquainted with the tree which
produced the quinquina, or Jesuits' bark, Plumier was
again prevailed upon to undertake a voyage to Peru, as
no one was thought so likely to investigate such a matter
satisfactorily. But soon after arriving at the port where
he was to embark, he was seized with a pleurisy, which
carried him off. This was in 1704, so that he was in
his fifty-eighth year.

Although the published works of Plumier are rather
voluminous, they bear a very small proportion to the
mass of writings and drawings he left behind him.
His " Description of the Plants of America" appeared
in 1693, 1 vol. fol. with 108 plates ; the latter executed at the expense of the government. " Nova
Plantarum Americanarum Genera" (1 vol. 4to. 1703)
comprehends 106 genera of American plants, with descriptions of about 700 species, many of them figured,
with detailed dissections of structure. His " Traité des
Fougères de l'Amérique " (Paris, 1705) is a handsome
folio volume, written in Latin and French, and illustrated with 172 plates. His manuscripts and drawings,
which are carefully preserved, are of such extent, and
many of the latter so highly finished, that it has excited
the surprise of many how he could find leisure to prepare
them. They relate to almost every branch of natural
history. The drawings of American fishes are particularly numerous and accurate ; and some of them
having accidentally passed into the hands of Bloch, who

eulogises them in high terms, he did not fail to avail himself of them for his great work. Upwards of 300 of his drawings of plants and other objects, also came into the possession of sir Joseph Banks. But the great majority were retained in Paris, where they were formerly preserved in the Bibliothèque du Roi.

Tournefort has consecrated a genus of plants (*Plumeria*), to the memory of his indefatigable friend, who has perhaps made us acquainted with a greater number of new natural productions than almost any scientific traveller that can be named. — J. D.

POEY, PH. — *Entomology.*

An entomological traveller, who visited Cuba.

Centurie de Lépidoptères de l'Isle de Cuba ; contenant la Description et les Figures coloriées de Cent Espèces de Papillons nouveaux ou peu connus, souvent avec la Chénille, la Chrysalide, et plusieurs Details Microscopique. Paris, 8vo., 1832. The larva and pupa, when given, are done in outline, but the perfect insects are finely coloured. The work, we believe, was discontinued after the three or four first numbers.

POHL, C. E. — *Comparative Anatomy.*

Expositio generalis Anatomica Organi Auditus per Classes Animalium. 1 vol. 4to. plates. Vindob. 1818.

POLI, X. — *Malacology and Comparative Anatomy.*

A distinguished and celebrated comparative anatomist, author of the most elaborate work on the molluscous animals ever published. He was high in command of the Neapolitan royal artillery, and tutor to the late king of the Two Sicilies. Although thus attached to the court, and holding several important offices under his sovereign, he still found time to prosecute his great work, although it proceeded very slowly from the press.

Testacea Utriusque Siciliæ, eorumque Historia et Anatome. 2 vols. folio, with most accurate and well-engraved plates. Printed at Parma, from the Bodom press, 1791—95. A Supplement to this was subquently published by Stephanus delle Chiaje. Naples, 1833. Some few copies were coloured.

PREYSLER, J. D. — *Entomology.*

Werzeichness Boehmischer Insecten. Pragæ, 1799. 4to.

PRICHARD, DR. — *Physician.*

1. *The Natural History of Animalcules ;* containing Descriptions of all the known Species of Infusoria. London, 1826. 2 vols. 8vo.
2. *Researches* into the Physical History of Mankind. London. 2 vols. 8vo. with plates. This valuable work has reached a second edition.

PRUNNER, LEONARD DE. — *Entomology.*

Lepidoptera Pedemontana. Turin, 1798. 8vo. This is an incomplete work ; the second volume never having appeared. It is very interesting, as the arrangement of the species is partly founded upon the structure of the larvæ.

RAFFLES, SIR THOMAS STAMFORD. — *General Zoology.*

It often happens with those who are most disposed to promote the interests of natural science, that their zeal is damped, and their exertions circumscribed, by having been placed by fortune in circumstances most unfavourable for giving effect to their wishes. Those, on the other hand, who occupy influential stations in foreign countries, where new objects of interest are most likely to be obtained, are, commonly, too much engrossed with the immediate duties of their office, or too indifferent to the subject, to avail themselves of the advantages within their reach.   In the case of sir Thomas Stamford Raffles, zeal and opportunity were happily combined; and we are, in consequence, indebted to him for most important additions to our knowledge of the zoology of a most interesting portion of the globe — the great islands of the East Indian Archipelago.

His father was captain of a vessel in the West India trade; and the subject of this notice was born at sea, off the island of Jamaica, on the 5th of July, 1781.   He was educated in an academy near Hammersmith, under the charge of Dr. Anderson; and at the early age of fourteen, became a clerk in the East India House. Aware of the defects in his education, he laboured with the utmost assiduity to repair them; and his studious habits, in connection with the duties of his clerkship, considerably impaired his health, which was at no time robust.   This, however, was restored by a pedestrian excursion through Wales : and in 1805, through the interest of a friend who had remarked in him qualities which deserved to be fostered, he was appointed assistant secretary to an establishment which the East India directors were then sending out to Penang.   He landed on the September following; and owing to the illness of the principal secretary, had all the duties of that office to perform from the outset, — a task for which he had well prepared himself by acquiring the Malay language during his outward voyage.   A recurrence of bad health compelled him to visit Malacca; where he interested himself in the history and antiquities of the people, in connection with Mr. Marsden, who was then employed in investigating these subjects.   Here, also, he became acquainted with Dr. John Leyden.

Mr. Raffles exhibited so much diplomatic skill and judgment in assisting to arrange the expedition against Java, that, when that island was captured in 1811, lord Minto, the governor-general of India, appointed him lieutenant-governor of Java and its dependencies.   In the discharge of the duties of this most arduous and re-

sponsible station, he took up his residence at Buitenzorg,
the seat of government; and, as soon as his official en-
gagements would permit, turned his attention to natural
history and antiquities. In the former department,
Dr. Horsfield, and other scientific gentlemen, were la-
bouring with great zeal and success : the antiquities
were very little known, but possessed of great interest,
owing to the numerous remains of Brahminical struc-
tures throughout the island,—proving that a colony of
Hindoos had settled there at a very early period. In
connection with Dr. Horsfield, Mr. Raffles re-established
the Society of Arts at Batavia, which had been the first
literary society founded by Europeans in the East, and
took upon himself the duties of president; in which capacity he delivered an address every year, containing a
summary of what had been accomplished by the society,
and holding forth encouragements to further exertions.
It being uncertain how the island of Java might be ultimately disposed of, his kind patron, lord Minto, before
consigning the government of the East to his successor,
lord Moira, made some provision for Mr. Raffles, by
procuring for him the residency of Fort Marlborough,
which gave him the first rank at Bencoolen. About
this time, he was subjected to much vexation by certain
charges brought against him by general Gillespie, respecting some acts of his administration when governor
of Java; which, however, were ultimately allowed to
drop, without leaving any imputation affecting either his
character or abilities. This annoyance was succeeded
by severe bereavements, in the death of Mrs. Raffles and
lord Minto;—afflictions, which so materially affected his
health, that, when he was superseded in the government
by Mr. Tindal,—an event which occurred soon after,—
he resolved on visiting England, and arrived in London
on the 16th of July, 1816. During his residence in this
country, he gained the friendship of most of the leading scientific men of the day ; received the honour of
knighthood from the Prince Regent ; and had the title
of lieutenant-governor of Bencoolen conferred on him

by the directors of the East India Company, with expressions of their high approbation and esteem. Accompanied by his lady, whom he had married while in
England, he sailed for his new residence in November,
1817, and arrived safely at Bencoolen. It was while
here, that he made those numerous and laborious excursions which produced such plentiful fruits in new and
highly interesting accessions to natural history. His
discoveries, indeed, were so numerous, that we cannot,
in this place, mention even the most important. He
had a regular establishment of naturalists and draughts-
men continually employed in preparing specimens and
figures ; and such a multitude of objects had he collected, that he describes his house as being a " perfect
Noah's ark." In short, no means were omitted to procure every thing which these fruitful regions afforded,
which might be of interest to the naturalist ; and had
the materials he amassed reached Europe, they would
have borne ample testimony to his extraordinary zeal
and intelligence. But this, alas ! was not destined to
take place.

Several sickly seasons having occurred in succession,
sir Thomas lost three of his children, and several of his
most attached friends ; and the lives of all were in jeo-
pardy : he therefore resolved to return to England ;
and, with this view, embarked all his collections, &c.,
and set sail in a vessel named the Fame, on the 2d of
February, 1824. On the evening of that same day, the
vessel took fire, and every thing was lost ; those on
board with the utmost difficulty escaping with their
lives. Sir Stamford again embarked, and reached England in safety ; but his constitution was now a good deal
shattered, and he did not survive his return to his native country above two years, — his death having taken
place on the 5th of July, 1826. For some time before
his decease, he was actively employed in organising a
Zoological Society in London, first projected by him ;
and the triumphant success of that scheme, with which
all are familiar, reflects the highest honour on his zeal

for science, and his liberality towards her followers. —
J. D.

Some valuable papers in the Linnæan Transactions
are all the writings he has left, to testify his great accuracy as a zoologist.

RAFINESQUE, SCHMALTZ, C. S.—*General Zoology and
Botany.*

A most enthusiastic and persevering naturalist.   Of
German extraction, he was settled, for many years,
as a merchant in Sicily ; which island he left for America about the year 1818.   He was soon after chosen
Professor of Natural History and Botany in the Tran-
sylvania University, which he subsequently resigned ;
and is now, we believe, a lecturer and commercial naturalist in Philadelphia. A strong prejudice has existed
against his writings and discoveries, both in Europe
and America, from the circumstance of his frequently
describing his "new genera and species" from hear-
say ; and from the extreme shortness, and consequent
insufficiency, of nearly all his descriptions.   Nevertheless, he has discovered and described a great number of
new objects ; and, but from his still adhering to these
vicious defects, might have ranked much higher in the
scientific world.   His innumerable papers are scattered
in Transactions, Annals, circulars, periodical maga-
zines, and even in single sheets ; while the few detached
works he has written, are mostly pamphlets : hence we
find it impossible to enumerate more than the following.
He was the first who called the attention of the American conchologists to the innumerable variety of bivalve
shellfish, which swarm in all their rivers.

1. *Caratteri di alcuni nuovi Genera*, et nuove
Specie de' Animali e Piante, della Sicilia.   Palermo,
1810. With rude figures.

2. *Indice* d'Ittiologia Siciliana.   Thin 8vo. Mes-
sina, 1810.

3. *Principes* Fondamentaux de Sémiologie. 1 vol. 12mo. Palermo, 1814.

4. *Analyse* de Univers, ou Tableau de la Nature. 8vo. Paris, 1815.

5. *Icthyologia Ohiensis*, or Fishes inhabiting the River Ohio, and its Tributary Streams. 8vo. pp. 90. Lexington, 1820.

Those who have the pleasure of knowing Mr. Rafinesque, will not fail to recognise his portrait in Audubon's most laughable account of "the Eccentric Naturalist," Ornitholog. Biography, vol. ii.

RANDOHR, C. A. — *Entomology.*

1. *On the Digestive Organs* of Insects (in German). 4to. Halle, 1811.

2. *Materials* for a History of German Monoculi. 4to. Halle, 1805.

RANG, SANDER. — *Malacology.*

An officer of royal marines in the French navy. A distinguished and very acute malacologist.

1. *Manuel* dè l'Histoire Naturelle des Mollusques et de leurs Coquilles. Paris, 1829. 1 vol. 12mo. with 5 plates. This is the most valuable of all the French introductions to conchology; since it contains numerous descriptions, collected and original, of the testaceous animals.

2. *Monograph of the Aplicidæ*, in folio, with coloured plates. This expensive work we have not seen.

RANZINI, THE ABBÉ CAMILLO. — *General Zoology.*

One of the best zoologists of Italy: at present, Professor of Natural History at Bologna.

1. *Elements of Zoology* (in Italian). Bologna, 1819.

Of which thirteen volumes have already appeared, all
relating to quadrupeds and birds.

2. *Memoirs* on Natural History (in Italian). Bologna, 1820. 1 vol. 4to. These works we have not
yet seen.

RAPP, WILLIAM.— *Molluscous Animals.*

One of the Professors at Tubingen, who has written
On the Polypi in general, and the *Actiniæ* in particular. 4to. Wiemar, 1829.

RAY, JOHN.— *Zoology and Botany.*

John Ray, the coadjutor of Willughby, holds a distinguished place in the annals of science, for his piety,
learning, and love of natural history. He was born at
Black Notley, in Essex, on the 29th of November, 1628.
At a proper age, he was removed to Cambridge, being
intended for the church; where, in 1644, he was entered
at Catherine Hall. In a few years he acquired a high
reputation, both for his scholarship and general attainments; a reputation to which he was indebted for the
friendship of most of the eminent men then attending
the university, among whom we find the well-known
names of Isaac Barrow, Dr. Tenison (afterwards arch-
bishop of Canterbury), and Dr. Arrowsmith. He was
ordained on the 23d of December, 1660, by Dr. Sander-
son, bishop of Lincoln. He continued to be a Fellow of
Trinity College till 1662, when he was deprived of his
fellowship for nonconformity.

The study of plants first attracted his attention, and
it seems to have always continued his favourite study.
His first work was a catalogue of the plants growing
about Cambridge, published in 1660. He then engaged in the preparation of a work applicable to the
whole kingdom; and this rendered it necessary for him
to travel in various directions. In his expedition to
Scotland, he was accompanied by Willughby and Mr.

Skippon ; but not visiting the Alpine districts, he made
but few additions to his flora.   His active mind, however, did not lose the opportunity these expeditions afforded him of becoming acquainted with every thing of
interest relating to the history, antiquities, language,
manners, &c. of the districts he traversed ; of which we
have ample proof in his journals, published after his
death by his biographer, Dr. Derham, under the title of
Itineraries.   In the spring of 1663, he extended his
researches to the continent, in company with Willughby,
Mr. Nathaniel Bacon, and Mr. (afterwards sir Philip)
Skippon ; the two latter of whom were Ray's pupils.
The journey extended to Sicily and Malta, and the results were published in 1673.*   In 1667, his love for
botany carried him into Cornwall ; where, also, he directed his attention to birds and fishes, in order to assist
his patron, Willughby, then engaged on his great work
upon animals, with whom, at this period of his life, he
chiefly resided.   In 1667, he was elected into the Royal
Society, and afterwards became a considerable contributor
to their Transactions. Having acquired, in his numerous
botanising excursions, an extensive acquaintance with
the proverbial expressions, local words and idioms, which
prevailed in different parts of the country, he drew up a
list of the former, and published it in Cambridge in 1672;
and a collection of unusual or local English words appeared about the same time.   In the year just named,
Ray sustained an irreparable loss, by the death of his
friend and patron, Mr. Willughby.   He was appointed
one of the executors, charged with the education of two
infant sons left by Mr. Willughby, and with the revision
and publication of his numerous manuscripts on the
natural history of animals.   In order to execute the various duties, Ray took up his residence at Middleton
Hall, and laboured with the utmost assiduity.   For the
use of his pupils, he published a *Nomenclator Classicus,*
the object of which was to give a correct explanation of

* Observations, Topographical, Moral, and Physiological, made in a
Journey through Part of the Low Countries, Germany, Italy, and France.

such Greek and Latin terms as apply more particularly
to natural objects. The first of Willughby's works,
which he prepared for the press, was a history of birds,
written in Latin, and published in 1675. Ray afterwards prepared an English translation, with considerable
additions both to the text and plates, which appeared
about three years later.

On the death of Mr. Willughby's mother, Ray removed from Middleton Hall, and went to Sutton Co-
field ; but, soon after, he removed to Falborne Hall, in
the vicinity of his native place. This year (1678), as
we find from an entry in his diary, his mother died, at
the age of seventy-eight. This event induced him to
take up his residence at Black Notley, when he completed the numerous works which were subsequently
published. The *Methodus Plantarum Nova*, containing
a methodical arrangement of vegetables, appeared in
1682; the *Historia Plantarum*, Vol. I., in 1686, and the
last in 1704 ; *Fasciculus Stirpium Britannicarum*, in
1688; and *Synopsis Methodica Stirpium Britannicarum*,
in 1690. The latter work is characterised by one of
the most competent judges, sir J. E. Smith, as one of
the most perfect systematical and practical floras of
any country, that ever came under his observation.
The next of Willughby's works which he prepared for
press, was the *Historia Piscium*, which forms a folio
volume, with 188 plates ; and in order to complete, in
connection with these larger works, an entire view of
the animal kingdom, Ray published various synopses of
different tribes of animals, the materials for which were
no doubt derived from Willughby's notes. He published, likewise, several works on Continental botany ; the
last volume in which he was engaged, was an *Historia
Insectorum*, which had been commenced by Willughby.
This, however, he did not complete to his satisfaction;
for he sunk under the attack of various disorders on the
17th of January, 1705. A monument is erected to
his memory in the church at Black Notley ; and, to
testify respect for one of the fathers of natural history

in this country, a commemorative meeting of his admirers was held in London, on the 29th of November, 1828, the second centenary of his birthday.

In the preceding sketch, we have chiefly alluded to Ray's writings on natural history: he published several others, not exclusively confined to that subject, but of a devotional character. Of these, the best known, and most highly esteemed, is "The Wisdom of God, manifested in the Works of Creation." — J. D.

Synopsis Methodium Avium et Piscium, Opus posthumum quod vivus recensuit et perfecit ipse insignissimus Auctor. 8vo. Lond. 1713. pp. 198. pl. 2.

RÉAUMUR, RÉNÉ ANTOINE FERCHAULT DE.

Réné Antoine Ferchault de Réaumur, one of the most ingenious observers and accomplished natural philosophers of his age, was born at Rochelle in the year 1683. He studied under the Jesuits at Poitiers, and afterwards at Bourges, at first with the intention of becoming a lawyer; but as his fortune was such as to enable him to follow his own inclinations, and these tended strongly to the natural sciences, the latter were soon permitted to engross all his attention. He went to Paris at the age of twenty; and his connections in that capital being of high rank and influence, he obtained, through them, ready access to every means of promoting his studies. So early as 1708 he became a member of the Academy of Sciences, to which body he had previously read several memoirs on geometrical subjects. About this period, his attention was chiefly occupied by examining, with a view to their improvement, the various processes carried on in the arts and manufactures; and he effected many beneficial changes by a judicious application of physical and chemical principles. In the prosecution of these inquiries, he made a discovery of great national importance, an account of which he laid before the public in 1722, in a treatise entitled "Traité sur l'Art de convertir le Fer en Acier, et d'adoucir le Fer fondu." The value of this discovery may be inferred from the fact, that all the steel previously used in France was imported. In gratitude for the benefit thus conferred on the nation, the duke of Orleans bestowed on him a pension of 20,000 livres.

Réaumur's scheme for graduating the thermometer, which consists in dividing the interval between the extreme points of the freezing and boiling of water into eighty degrees, on the principle that spirits of wine, in a certain state of rectification, expand 80,000 parts, has been generally admitted to be a very simple and convenient one. It has been universally adopted in the South of Europe, but that of Fahrenheit seems to be preferred in the North.

Many subjects of natural history and physiology were at different times investigated by him, and always with a closeness of observation, conclusiveness of experiment, and pains-taking anxiety for the truth, which led to the most satisfactory results. He was the first naturalist who formed a zoological museum of any extent in France, and it afterwards formed the ground-work of the Royal Parisian Museum. During Réaumur's life-time, it was placed under the charge of Brisson.

But no subject engaged the attention of Réaumur more than insects ; and what he has done in this department, entitles him to the lasting gratitude of naturalists. His great work on this subject gives a most interesting detail of external, sometimes of internal, structure, manners, economy, &c. The whole is amply illustrated with plates, which, although not of much merit as works of art, are sufficiently well executed to render the text perfectly intelligible. In his anxiety to leave nothing to the reader to supply, he is rather apt to become diffuse and prolix ; and his utter disregard of system and nomenclature often renders it difficult to determine to what species his observations strictly apply. An occasional inconclusiveness in his experiments is also observable ; as, for example, when he infers that the larvæ of

carnivorous flies (such as *Sarcophaga*) will not consume
the flesh of a living animal, because, on placing some of
them beneath the skin of a domestic fowl, he found
them make no attempt to feed upon its substance. But
the work, notwithstanding these defects, is really valuable. Engrossed with such pursuits, Réaumur passed
a very tranquil life, sometimes residing on his estate in
Saintonge, at other times in his country-house near
Paris. His death took place on the 18th of October,
1757.

1. *Mémoires* pour servir à l'Histoire des Insectes.
Paris, 1734—42. 6 vols. 4to. Cuvier informs us
that a seventh volume exists in manuscript.

2. *Art de faire éclore* et d'élever en toute Saison des
Oiseaux domestiques de toutes Espèces. Paris, 1751.
2 vols. 12mo. Plates in vol. i., 9.: in vol. ii., 7.

REDI, FRANCESCO.

A learned physician of the seventeenth century. He
enjoyed much celebrity as a literary character, and for his
curious, although cruel, experiments upon animals.

1. *Experimenta* circa Generationem Insectorum.
Amstelodami, 1686. 3 vols. 12mo.

2. *Opuscula Varia.* Lugd. Bat. 1729. 3 vols.
18mo. Plates, vol. i., 31.; vol. ii., 28.; vol. iii., 26.

REGENFUSS, F. M. — *Conchology.*

Choix de Coquillages, de Limaçons et de Crustacés,
with 12 engravings representing 78 shells, and with
a portrait of Frederick V. Cop. 1758. Royal folio.
This excessively rare work was privately printed at
the expense of the monarch whose portrait forms
the frontispiece.

REICHENBACH, H. T. L. — *Entomology.*

Monographia Pselaphorum. Lipsiæ, 1816. 1 vol.
8vo. plates.

x 2

REINWARDT.—*Zoological Traveller. General Zoology.*

One of the chief professors at Leyden, who travelled through the Indian islands, particularly Java, where he made splendid collections. We frequently see his name attached to species ; but we cannot learn what he has written. Cuvier is also silent upon this point.

RÉNARD, LOUIS.—*Ichthyology.*

Poissons, Ecrévisses et Crabes, que l'on trouve autour des Isles Moluques, et sur les Côtes des Terres Australes. Amsterdam, 1754. 2 parts, 1 vol. folio. pl. col. 100. This is a very curious collection of plates, engraved from drawings of native artists. They are obviously done by different hands, and therefore of unequal merit. Many show indications of being greatly exaggerated ; yet a large proportion, judging from the degree of similitude with which some of the known species are delineated, are, no doubt, tolerably faithful.

RETZIUS, PROFESSOR. — *Botanist and Zoologist.*

Besides many essays and papers in Transactions, he published the
Fauna Suecica of Linnæus, greatly enlarged.

RICHARDSON, JOHN. — *Zoology and Botany.*

Celebrated as the companion of captain Franklin in his perilous voyage and journey in the Arctic regions; and chief surgeon and naturalist to that expedition. On his return, he was commissioned to arrange the zoological discoveries for publication ; and is now holding a high medical office in the Navy Hospital.

1. *Appendix to Captain Franklin's Voyage,* con-

taining the Botany and Zoology. London, 1823.
1 vol. 4to.

2. *Fauna Boreali Americana* (generally cited as
Northern Zoology), or the Zoology of the Northern
Parts of British America, collected on the Northern
Land Expeditions, under the Command of Sir John
Franklin. London, 1828, &c. 4 vols. 4to. with
numerous plates. These volumes form as many distinct works; viz. 1. The Quadrupeds, by Richardson;
— 2. The Birds, by Richardson and Swainson;—
3. The Insects, by Kirby;—and, 4. The Fishes, by
Richardson. He has likewise written several zoological papers, inserted in Transactions; particularly
one, in 1840, on the new fish of the Australian Seas.

RIDINGER, J. E.—*Mammalogy.*

Betrachtung der Wilden Thiere, mit beygefügter
vortrefflichen Poesie des Herrn Barth. Heinr. Broches.
Augsburgh, 1736. folio.

RISSO, A.—*Entomology and Ichthyology.*

An apothecary of Nice; the marine animals of which
place he has investigated with great zeal and ability.

1. *Ichthyologie de Nice,* ou Histoire Naturelle des
Poissons du Département des Alpes Maritimes. Paris,
1810. 8vo. pp. 388. pl. col. 11. Figures and descriptions of numerous rare and new species are contained in this volume.

2. *Histoire Naturelle des Crustacés* des Environs
de Nice. Paris, 1816. 1 vol. 8vo. pl.

ROBINEAUX, DESVOIDY.—*Entomology.*

A French physician, who has chiefly written various
papers on the *Diptera.* In these he proposes upwards
of 300 new generic names for the single family of
Muscida. The spirit in which this is done, may be

x 3

judged of by the fact of his having made no allusion to the admirable work of Meigen! and scarcely to any modern author. It is hardly necessary to say that the characters of these proposed groups are generally too trivial to deserve notice, and the whole work is undeserving of authority.

ROCHEFORT, N. — *Zoology.*

A protestant minister of Holland, who seems to have resided for some years in the West Indies.

Histoire Naturelle et Morale des Antilles de l'Amérique. Rot. 1658. 1 vol. 8vo. This edition appeared without the name of the author.

RŒMER, J. C. — *Entomology.*

Genera Insectorum Linnæi et Fabricii, Iconibus illustrata. Vitoduri Helvetiorum, 1789. 1 vol. 4to. The text is meagre; but the plates — admirably designed, and correctly coloured — contain a great number of figures, and render the volume a valuable work of reference.

RŒSEL DE ROSENHOF, A. J. — *Entomology.*

Born in 1705. An ingenious observer of nature, and one of the most eminent zoological artists that have yet appeared in Germany. He resided at Nuremberg, and died in 1759.

1. *Historia Naturalis Ranarum* nostratium. Nuremb. 1758. folio.

2. Der Monatlich-herausgegebenen Insecten Belustigung, i. e. a Monthly Publication of the Amusements of Insects. Nuremberg, 1746. 4 vols. 4to. The plates are numerous, and are designed and executed with admirable truth and delicacy. A fifth or

supplementary volume to this valuable work was
published by Kleemann, in 1761.

ROISSY, FELIX DE. — *General Zoology.*

His name occurs as author of the conchological portion (in conjunction with De Montfort) of the edition
of Buffon edited by Sonnini.

RONDELET, GUILLAUME. — *Ichthyology.*

Born in 1507; became Professor of Medicine at
Montpellier; and died in 1566.

Gulielmi Rondeletii Libri de Piscibus Marinis, in
quibus veræ Piscium Effigies expressæ sunt. Lugduni, 1554. 1 vol. folio. One of the very few books
upon natural history, published at that remote period,
which is useful in modern times. The figures on
wood, although in some respects rude, when viewed
as sketches are admirable; and many of them are
copied into the hot-pressed publications on natural history, got up by the booksellers.

ROSSI, PIETRO. — *Entomology.*

Professor of Natural History at Pisa.

1. *Fauna Etrusca,* sistens Insecta quæ in Pro-
vinciis Florentinâ et Pisanâ præsertim collegit Petrus
Rossius. Liburni, 1790. 2 vols. 4to. col. pl. 10. The
figures are few, but tolerably accurate. There is another edition, in 8vo., printed at Helmstad: the first
volume in 1795; the second in 1807.

2. *Mantissa Insectorum,* exhibens Species nuper
in Etruscâ collectas, a Petro Rossio. Pisis, 1792—
1794. 2 vols. 4to. col. pl.

Roux, Polydore. — *General Zoology.*

An able zoologist of Marseilles, and curator of the cabinet of Natural History in that city.

1. *Ornithologie Provençale,* ou Déscription, avec Figures coloriées, de tous les Oiseaux qui habitent constamment la Provence, ou qui n'y sont que de Passage ; suivie d'un Abrégé des Classes, &c. Small 4to. Paris, published in numbers, with 8 coloured lithographic plates in each, of which we have seen 49. The figures are of similar execution to those of the generality of French works, but the lithographic printing is not very good.

2. *Crustacés* de la Méditerranée, et de son Littoral. Uniform with the above, and proposed to be completed in fifty-six numbers, but of which we have only seen three. The plates are coloured, and remarkably good ; but we are fearful this excellent work has been discontinued.

3. *Iconographie Conchyliogique,* ou Recueil de Planches lithographiées et coloriées réprésentant les Coquilles Marines, Terrestres, et Fossiles, &c. 4to. Only one number is noticed in the booksellers' catalogues.

Rozet, M.

Captain in the general staff, and Member of the Natural History Society of Paris.

Voyage dans la Régence d'Alger, ou Description du Pays occupé par l'Armée Française en Afrique; contenant des Observations sur l'Histoire Naturelle, &c. Paris, 1833. 3 vols. 8vo. and an Atlas of plates. This work is stated to contain many new subjects of zoology, particularly among the reptiles.

Rudolphi, C. A. — *Intestinal Worms.*

Professór at Gripswald, &c.; subsequently at Berlin:

an acute comparative anatomist, and our principal authority on the intestinal worms.

    Entozooa, seu Vermium Intestinalium Historia
Naturalis. Amst. 1808. 2 vols. 8vo.

RUMPH, G. E., OR RUMPHIUS. — *Conchology and Botany.*

    Born at Hanau, in 1637.   He settled in Holland as
an East India merchant, and acquired great wealth and
reputation.   At a subsequent period, he was made governor of Amboyna, where he resided several years,
devoting all his leisure to investigating the botany and
conchology of that fruitful island.   On his return to
Europe, he published his discoveries in both these departments of nature, and devoted his time and fortune
to the formation of a superb museum.   Dr. Turton
mentions (but without citing his authority), that the
first specimen brought to Holland of the beautiful Venus
Dione (now a common shell), was actually purchased
by Rumphius for a sum equal to 1000*l*. !   Such enormous folly, however, is scarcely credible.   Towards the
latter period of his life, Rumphius became blind; yet still
(like the late lord Coventry), this heavy affliction did
not abate his love for nature.   His portrait is affixed to
his first work : it represents a venerable and peculiarly
expressive countenance; a tray of shells is on the table,
and the hands seemingly employed in fulfilling the office
of the eyes.   Rumphius expired at the age of sixty-seven,
in the year 1706.

    1. *Cabinet of Amboyna* (in Dutch).   Amsterdam,
1705. folio.

    2. *Thesaurus Imaginum Piscium Testaceorum,*
&c.   Hagæ Comitum, 1739. folio, pl. 60.   The
same plates were used for both these works; but the
text of the latter is much more scanty.

RÜPPELL, DR. EDWARD.—*General Zoology.*

One of the most celebrated zoological travellers of the
present day, whose researches in Northern Africa and
the Red Sea have enriched science with vast additions.
His chief works are the two following, which are exclusive of several smaller essays, &c. on ichthyology and
ornithology.

    1. *Atlas zu Reise im Nördlichen Africa,* von
Ed. Rüppell. Frankfurd, 1128. The quadrupeds,
birds, fish, &c. are arranged, and may be bound separately. The plates are drawn on stone, well coloured, and appear very accurate. Small folio.

    2. *Neue Wirbelthiere* zu der Fauna von Abyssinien
Gehörig. Uniform with the above. Frankf. 1838.
Small folio.

RUSSELL, DR. A.—*Oriental Traveller.*

Natural History of Aleppo, containing a Description
of the City, and the principal Natural Productions of
its Neighbourhood. 2 vols. royal 4to. with several
plates, and much information on zoology. A second
edition was published by his son, Dr. Patrick Russell.
London, 1794.

RUSSELL, PATRICK.—*Zoology and Botany.*

Botanist and Naturalist to the East India Company,
whose works are valuable for the numerous figures and
descriptions they contain of animals collected by him in
India.

    1. *Account of Indian Serpents,* collected on the
coast of Coromandel. 2 vols. folio, 92 coloured plates.
London, 1796—1800.

    2. *Fishes;* Descriptions and figures of 200 species,
collected at Vizagapatam. London, 1803.

SALERNE. — *Ornithology.*

Was a physician of Orleans.

L'Histoire éclaircie dans une de ses Parties principales, &c.   Traduite du Latin du Synopse de Ray,
augmentée par M. Salerne.   Paris, 1767. 4to. pp.
467. pl. 31.   Cuvier remarks, that the figures are by
the same artist (Martinet) who designed those for
the Planches enluminées of Buffon, and the Ornithologie of Brisson, and that they are mostly taken
from the same specimens.

SALVIANI, HIPPOLITO. — *Ichthyology.*

This eminent naturalist was born, it is said, of noble
parentage, in the year 1514, in *La Citta di Castello,* on
the Tiber, the ancient *Tifornum Mitaurense,* in the
duchy of Urbino.   It does not appear where or how his
early education was conducted.   He, however, studied
medicine ; and after visiting several of the towns of
Italy, settled as physician at Rome.   Here he became
eminent in the teaching as well as the practice of the
seience ; being appointed ordinary physician successively
to popes Julius III., Marcelus II., and Paul IV. ; and
teaching with much celebrity, we believe, for 22 years.
To his more strictly professional avocations, he conjoined
the kindred one of zoology; and soon became as famous
as a naturalist, as he was as a physician.   His love for
the science led him to investigate it not merely in books,
but in nature; being anxious to examine every thing
curious in this department which found its way to the
Great City ; and always, when he could, procuring it
for his museum.   He seems early to have selected
ichthyology as the object of his peculiar study; and
by his active energy at home, and his extensive correspondence abroad, soon discovered many new species,
as well as illustrated others which were but obscurely
known.   Thus, his friend Ghinius sent him to Rome a

drawing of that extraordinary looking fish, the shark
sun-fish of British authors, which, at the moment, was
regarded as an experiment upon his own credulity and
that of others ; and great, therefore, was the satisfaction
experienced, when one of these fishes was soon afterwards
brought to the Roman market, and became the subject
of minute investigation.

His stock of information thus accumulating, he was
led to its publication ; and in a work which, whether we
regard its dimensions, or printing, or engravings, will
not be considered less magnificent now, than nearly
300 years ago.   Nor was its value confined to these
adventitious helps.   The principle upon which it was
composed, was new at the time, and was a good one,
and the execution did honour to the plan.   " There are
many," he remarks, " who transfer what they read in
others to their own works, without considering whether
the statements are true or false ; rather following the
authority of man than the truth of history.   I have
been ever determined, however, on the contrary, to state
nothing, the truth of which I have not ascertained." *
The great work, which will immortalise his name, was
printed in his own mansion, and, with the engraving of
the copper-plates, must have occasioned much trouble
and expense.   These engravings would do honour to the
arts at the present day ; and hence, as well as from the
mass of ancient lore it contains, the high value that is
now attached to the volume.   Salviani wrote other
works,— some on physic, and others on lighter subjects,
— several of which came to a second edition.   He left
two sons ; one of whom, under his father's auspices, became physician in Rome.   He died in the year 1572.
— J. D.

Aquatilium Animalium Historia.   Romæ, 1554.
folio. pp. 256. pl. 81. The figures are numerous, and,
for the time in which they were executed, very good.

* From a memoir, in a volume of the "Naturalist's Library," just published.—J. D.

SAMOUELLE, GEORGE.

One of the officers of the British Museum.   An assiduous and accurate observer of nature ; whose work
has done more to advance the study of British insects,
upon improved and enlarged views, than any other in
the English language.

The Entomologist's Useful Compendium, or an Introduction to the Knowledge of British Insects.
Lond. 1819. 8vo. pp. 496. pl. 12.

SAVIGNY, JULES CÉSAR. — *General Zoology.*

A celebrated naturalist, and one of the most profound
observers of nature.   The investigations of Savigny
have been few; but whatever subject he has touched, it
has been with such consummate talent, as to leave
nothing to be done by those who might follow him.   He
was the first who discovered the existence of rudimentary
jaws in the lepidopterous insects ; while his inimitable
dissections of the *Tunicata,* and other invertebrate
animals, will remain a lasting monument of his skill and
perseverance.   His constant use of the microscope, unfortunately, produced blindness.   But the French government, sensible of his talents, have assigned him a
handsome pension.

1. *Hist. Nat.* et Mythologique de l'Ibis.   Paris,
1805. 8vo.

2. *Mémoires* sur les Animaux sans Vertèbres.
Paris, 1816.   2 parts, 8vo. Pl. in pt. i., 8.; in pt. ii.,
pl. 24.

3. *Mémoires sur les Oiseaux* de l'Egypte; inserted
in the grand work on that country, published under
the patronage of the emperor Napoleon.

SAY, THOMAS. — *General Zoology.*

Born in 1787; died in 1834.   One of the best known
zoologists of America, whose numerous writings, almost

confined to Transactions and periodicals, have greatly illustrated the zoology of that country.

1. *American Entomology*, with coloured plates. 3 vols. 8vo. 1824—1826. So rare are the American works on Zoology in England, that we have only seen one copy of the first volume.

SCATTAGLIA. — *Mammalogy.*

Descrizione degli Animali Quadrupedi. Venezia, 1771. 4 vols. folio, 200 col. pl.

SCHÆFFER, JACQUES CHRETIEN. — *Entomology.*

He was a clergyman at Ratisbon, the natural productions of which place he investigated with great zeal and success. In entomology, he proposed several good genera; but his works are now only valuable on account of their figures; these are numerous, and tolerably well executed for the period. He was born in 1718, and died in 1790.

1. *Elementa Ornithologica* Iconibus vivis Coloribus expressis illustrata. Ratisbonæ, 1764. 4to. pl. col. 70.

2. *Museum Ornithologicum*, exhibens Enumerationem et Descriptionem Avium. Ratisbon, 1789. 4to. pp. 72. pl. col. 52.

3. *Elementa Entomologica*. Regensburg, 1766. 1 vol. 4to.

4. *Icones Insectorum* circa Ratisbonam indigenorum. Regensburg, 1769. 3 vols. 4to. Each volume contains plates; two being printed upon one leaf. No arrangement is adopted respecting the subjects, but they are very numerous, and in general easily recognised.

5. *Apus pesciformis* (Cancer stagnalis *Linn.*), Insecti Aquatici Species noviter detecta. Ratisbonæ, 1759. 4to. coloured plates.

6. *Monograph of the Genus Apus.* 1 vol. 4to. coloured plates.

7. *Abhandlungen von Insecten.* Regensburg, 1764 —1779. 1 vol. 4to.

SCHÆPF, J. D.—*Erpetology.*

Physician of Anspach.   Born in 1752.

Historia Testudinum Iconibus illustrata.   Erlang, 1792, &c. 4to. pl. col.

SCHELLENBERG, J. R.—*Entomology.*

An engraver and artist of Zurich.

1. *Cimicum* in Helvetiæ Aquis et Terris degentium Genus Turici. 8vo. 1 vol. 14 plates.

2. *Genres des Mouches Diptères.*   Zurich, 1803. 1 vol. 8vo. with 42 coloured plates.   The descriptions are in French and German, and are done by other hands.

SCHEUCHZER, J. J.—*Erpetology.*

Physician at Zurich.   Born in 1672; died in 1733.

Dissert. Physica Sacra. Amst. 1732. 4 vols. folio. Cuvier observes, that this book is interesting to the naturalist, from the numerous figures of serpents it contains.   There are near 700 plates, some of which are surrounded with figures of animals.

SCHNEIDER, J. G.—*Ichthyology.*

A celebrated Hellenist, who was Professor of Natural History at Francfort, and afterwards at Breslau.

1. *Histoire Naturelle* générale des Tortues (in German).   Leipsig, 1783. 1 vol. 8vo.

2. *Amphibiorum* Physiologiæ Specimen 1 et 2. Zullichaw, 1797. 2 parts, 4to.

3. *Historia Amphibiorum* Naturalis et Liteariæ. Fasc. 1. et 2. Jena, 1799—1801. 8vo.

Schneider also edited the *Systema Ichthyologiæ* of Bloch. Berlin, 1801. 2 vols. 8vo. with 110 plates.

SCHŒNHERR, C. J. — *Entomology.*

1. *Synonymia Insectorum.* Stockh. 1806—1808. 2 vols. 8vo. plates. An Appendix was published in 1817.

2. *Curculionum* disposito Methodica, seu Prodromus ad Synonymiæ Insectorum, &c. Lipsiæ, 1826. 1 vol. 8vo.

SCHONEFELD, E. DE. — *Ichthyology.*

Physician at Hamburgh.

Ichthyologia Ducatuum Slesvigi et Holsatiæ. Hamb. 1624. 1 vol. 4to.

SCHRANK, F. P. — *Entomology.*

Professor of Natural History at Ingolstadt. Born in ——; died in 1747. His entomological writings are valuable, although chiefly scattered in periodicals.

1. *Enumeratio* Insectorum Austriæ indigenorum. Aug. Vindelic. 1781. 1 vol. 8vo. plates.

2. *Fauna Boica.* Nuremb. 1798, &c. 6 vols. 8vo.

SCHREBER, J. CH. D. — *Mammalogy.*

Professor at Erlang. Born in 1739.

Naturgeschichte der Säugthiere, &c.; or, The Natural History of Quadrupeds. 4to. coloured plates. This work has been published at uncertain intervals from 1775 to 1818. Wood informs us that its continuation may be expected from Dr. Goldfüss, of Erlang, but that its progress is slow and uncertain.

The greatest number of the figures are copied from those of Buffon, and evidently coloured from descriptions : some few are original, but drawn without scientific knowledge or good taste. The five first parts contain 347 plates.

SCHRŒTER, J. S. — *Entomology and Conchology.*

A protestant minister of Buttstedt, and author of many works on conchology : they are but little quoted, except for their figures, being written entirely in the German language. He was born in 1735; died in——.

1. *Ueber den innern* Baw der See, &c.; or, An Account of the Internal Structure of Sea Shells, and some foreign Land and River Spirals. Frankfort, 1783. 4to. pp. 164. pl. 5. The plates represent the sections of shells.

2. *Einleitung* in die Conchylienkentniss nach Linné, &c.; or, An Introduction to the Linnæan System of Conchology. Halle, 1783—1786. 3 vols. 8vo. pl. 9.

3. *Der Geschichte* der Flüsconchilien; or, A Treatise on River Shells. Halle, 1779. 4to. pp. 434. pl. 11. Of this work, according to Wood, the nine first plates are coloured, but the two last are always plain.

Percheron enumerates more than 20 entomological papers by this writer, scattered in periodicals.

SCILLA, AUGUSTINO. — *Malacology.*

An ingenious Sicilian, who first pointed out the analogy between fossil and recent bodies. The title of this book is curious.

1. *La Vana Speculatione* disingannata dal Senso. Napoli, 1670. small 4to.

2. *De Corporibus Marinis* Lapidicentibus. Romæ, 1752. 4to. pl. 28. This is a Latin translation of the above; the plates are very good.

Schumacher, C. F. — *Intestinal Worms.*

Essai d'un Nouveau Système des Habitations des Vers Testacé. Copenhagen, 1818. 4to. pp. 287. pl. 22.

Scopoli, J. Ant. — *Entomology.*

A celebrated naturalist and physician of the last century. He was born at Cavalese, in 1725; and after completing his studies at Venice, he traversed the mountains of Carneola and the Tyrol : on his return, he published his botanical and entomological discoveries. In 1754, he went to Venice ; and soon after was nominated physician in chief to the Austrian mining establishment in Tyrol. Here he remained near ten years ; till at length, after much solicitation, he was created a counsellor of mines, and Professor of Mineralogy at Schemitz. He laboured in these offices with much assiduity; yet it was only after another interval of ten years that he obtained the professorship of botany and chemistry at Padua. Here he published several medical essays; and described the contents of the University Museum in his *Deliciæ,* &c. His domestic life is said to have been embittered by many afflictions, and his public conduct subjected to unmerited censure. Scopoli was one of those few members of the university, who had the courage and honesty to expose the disgraceful conduct of the celebrated Spallanzani, whom they detected in robbing the public museum. The talents of the culprit supported him, however, in the good opinion of the emperor, who merely bestowed upon him a rebuke ; and then silenced his accusers in such a way, that the mortification, joined to the debilitated state of his health, caused the death of Scopoli. For more than a year previously, by too much application to the microscope, he had lost his sight. He expired at Padua on the 3d of May, 1788. His descriptions are not now of much value.

1. *Introductio* ad Historiam Naturalem.   Prague,
1777. 1 vol. 8vo.
2. *Anni* Historico-Naturales.   Lipsiæ, 1768—
1772. 1 vol. 8vo.
3. *Entomologia* Carniolica.   Vindebonæ, 1763.
1 vol. 8vo.
4. *Deliciæ* Floræ et Faunæ Insubricæ.   Ticini,
1768—1788. 1 vol. folio.

SEBA, ALBERTUS. — *General Zoology.*

A wealthy apothecary of Amsterdam. Seba was born
in 1663, and seems to have had a passionate love for
collecting ; but he was entirely destitute of any scientific
or critical knowledge of his possessions. Hence, the
text of his large work, which must have cost him a
great sum, is almost useless; while the majority of the
plates deserve great praise, and are quoted with confidence. The figures of shells are much superior, both
in design and execution, to those of Favanne, Martini,
Gualter, or any of his predecessors; but, unfortunately,
most of them are reversed, by the ignorance of the engraver. He only lived to see the first volume published,
as he died in 1736, two years after its appearance.

Locupletissimi Rerum Naturalium Thesauri ac-
curata Descriptio. Amsterdami, 1734—1765. 4 vols.
folio. A few copies were coloured in a very superior
manner.   The plates, divested of their obsolete descriptions, have more recently been published under
the following title : —

Planches de Seba, accompagnées d'un Texte ex-
planatif mis au courant de la Science, et rédigé par
un Réunion de Savans. 30 folio parts, at 4s. each.
Paris, ab. 1835.

SELBY, JOHN PRIDEAUX. — *Ornithology.*

Mr. Selby is the exclusive author of the most splendid
and costly work yet published on the birds of Great

Y 2

Britain, viz. *Illustrations of British Ornithology,* ele-
phant folio.   Most of the figures are of the natural
size.   The land birds form seven parts ; and the water,
eleven.   Edin. 1821—34.   The letter press is contained
in two octavo volumes.   It is sold separately, and is
one of the best works extant on our native ornithology.

SENGUERDIUS, WOLFERD.—*Entomology.*

Tractatus Physicus de Tarantulâ.   Lugduni Batavorum. 1 vol. 12mo.

SERRES, MARCEL DE.—*Physiologist.*

A learned comparative anatomist, distinguished by his
writings on the internal structure of insects.   He is Professor of Mineralogy at Montpellier.

Mémoire sur les Yeux composés et les Yeux lisses
des Insectes.   Montpellier, 1813. 1 vol. 8vo. with
figures.

SERVILLE, J. G. AUDINET.—*Entomology.*

An able entomologist, whose genera, nevertheless,
appear to us, in numerous instances, too trivial to be
adopted.

1. *Orthoptères,* Revue Méthodique des Insectes de
l'Ordre des.   Paris, 1831. 8vo.
2. *Nouvelle Classification* de la Famille des Longi-
cornes.   A valuable paper, published in the Annales
de la Soc. Ent. de France.   1832—35.

SHAW, DR. GEORGE.—*General Zoology.*

The predecessor of Dr. Leach in the British Museum.
A laborious writer and compiler, whose works are now
of little value.   The chief of these are : —

1. *General Zoology,* or Systematic Natural History,

continued by Stevens. 28 parts, forming 14 vols., with numerous plates. 8vo. London, 1800—27. Nearly the whole compiled from other authors.

2. *Naturalist's Miscellany.* 24 vols. royal 8vo. with 1068 coloured plates. With the exception of a very few, the whole of these plates are indifferent copies, taken out of other works, and often coloured from descriptions.

3. *Zoological Lectures.* 2 vols. 8vo. London, 1809.

4. *Museum Leverianum*, with coloured figures of Birds, &c., from the Museum of the late Sir Ashton Lever. 1 vol. 4to. 1792. The plates, as the title implies, are all original.

SHUCKARD, WILL. ED.—*Entomology.*

Librarian to the Royal Society of London. One of the most luminous and philosophic entomologists of the present day. He has more especially devoted his talents to the investigation of the *Hymenoptera,*—an order upon which he is considered the first authority in this country.

1. *Fossorial Hymenoptera,* Essay on the Indigenous; containing a Description of all the British Species of burrowing Sand-wasps. 1 vol. 8vo. with plates. London. 1837. We have been very much struck with this lucid and masterly treatise. It breathes throughout not only a philosophic but a candid mind, and shows the writer to be far more than a mere nomenclator of genera and species. Although he very properly points out the errors of others, he yet does ample justice to their merits on every occasion. It is delightful to trace this spirit, so different from that *littleness* which leads to a style of writing totally different. Mr. Shuckard's works, more especially this and the next, are essential to every British entomologist, and may be studied with great advantage by the most eminent. Another volume, *On the British Bees, Ants, and Wasps*, has engaged the learned

author's attention for many years : it will be a fit
companion to the above, and is looked for with much
interest by the entomological public.

2. *The British Coleoptera* delineated. 1 vol. 8vo.
London, 1840. This is one of the most valuable
illustrated works, and certainly the cheapest, that the
student can possess. (See SPRY.)

3. *Elements* of British Entomology. Part I. We
regret not having seen this work.

4. *Manual of Entomology*, translated from the
German of Burmerster. 1 vol. 8vo. London, 1837.
With plates.

Mr. Shuckard has also written several interesting
papers in periodicals. In Taylor's Annals of Natural
History, is a monograph of the *Dorylidæ;* and in the
Entomological Trans., is an elaborate paper on the wings
of the hymenopterous order, full of original views, and
deserving the greatest attention.

SLABBER, MARTIN.

Recreations from Nature, contained in Observations
with the Microscope (in Dutch). Harlem, 1778. 4to.

SIR HANS SLOANE. — *General Zoology.*

This distinguished physician, and princely patron of
the sciences, to whom the nation is indebted for the
foundation of our national museum, was born at Killi-
leigh, in Ireland, on the 16th of April, 1660. His father,
Alexander Sloane, was a native of Scotland, but had
settled in Ireland along with the colony of Scotchmen
sent thither by James I., and in which country the
earlier portion of his life was spent. It seems probable
that the natural delicacy of his constitution, which prevented him from engaging in any very laborious pursuits,
induced him to turn his attention to the study of nature.
Having resolved to follow the medical profession, he
repaired to London to prosecute his studies ; and after

continuing to do so for several years with great diligence
and success, he visited the Continent in search of further
instruction. While there, he gratified his taste for botany,—a part of natural history for which he had always
a very strong predilection,— by attending the lectures of
the celebrated Tournefort; and also those of M. Magnol
at Montpellier. At this place, also, he is supposed to
have taken his degree of M.D. He returned to England
in the close of 1684.

He now became domesticated in the family of the
well-known Dr. Sydenham, and began to practise as a
physician. On the 26th of November, 1684, he was
proposed by Dr. Martin Lister as a candidate for mem-
bership in the Royal Society, and was elected on the
21st of January following. His election as Fellow of
the College of Physicians took place on the 12th of April,
1687. On the 12th of September of the same year, he
set sail for Jamaica, in the capacity of physician to the
duke of Albemarle, who had been appointed governor of
that island, the great inducement to this step being the
opportunity it afforded of examining the natural productions of that country. During his brief stay,—for he
remained only fifteen months,— he was indefatigable in
collecting plants and other natural objects; and he even
attempted, but without success, to bring several live
animals home with him. The materials thus amassed,
were rendered available to science by the publication,
first, in 1696, of a " Catalogue of Jamaica Plants;" and
secondly, a " Natural History of Jamaica,"—a splendid
folio, illustrated in the best style the arts then admitted
of. The first volume was dedicated to queen Anne ; the
second (which did not appear till 1725), to George I.
On his return from Jamaica, his practice as a physician
became great; but his scientific reputation even surpassed his professional fame. In 1693, he was elected
secretary to the Royal Society ; an office which he held
with high distinction till 1712. Professional distinctions
were likewise showered upon him at various times : in
1694, he was chosen physician of Christ's Hospital; and

at a later period (1719), President of the College of
Physicians. On the accession of George I. (1716), he
was created a baronet,—a title which had never been
conferred on an English physician before his time; and
he was soon after appointed physician-general to the
army.   These various appointments, in connection with
an extensive and lucrative practice, enabled him to acquire a very considerable fortune, which he expended
with the utmost liberality, both for the promotion of
science and for charitable purposes.   Scarcely any of
the London charities or medical establishments failed to
experience his bounty.   The Company of Apothecaries,
in gratitude for the signal favours he conferred on them,
erected a marble statue of him (by Rysbach), in the
centre of their botanical garden.*   Upon the death of
sir Isaac Newton, in 1727, sir Hans' celebrity, position
in society, and extensive influence, naturally pointed
him out as the most worthy successor to the presidency
of the Royal Society : he was accordingly chosen for
that distinguished station, and continued to hold it, to
the great advantage of the society, for 13 years, when,
from his great age, he retired from public life.

The nucleus of his museum was the collection of natural objects made in Jamaica, of which an account will
be found in *Evelyn's Diary*, under the date of April
16. 1691.   To these he never failed to add whatever
his ample means could purchase ; and his numerous
friends were continually contributing objects of interest
and value.   His collection received a considerable increase, in 1702, by Mr. Courten bequeathing to him the
whole of his museum and curiosities ; and another important addition was made by the purchase of Petiver's.
In January, 1741, he removed his library and collections
from his house in Bloomsbury to that at Chelsea, which
was fitted up in such a manner as to display the objects
to advantage ; and here he spent the remainder of his

* The inscription is as follows : — Hansio Sloane, Baronetto. Achiatro
Insignissimo Botanices Fautori, Hoc honoris causâ Monumentum Inque
perpetuum ejus Memoriam Sacrum, voluit Societas Pharmacopeiorum,
Londinensis, 1733.

days among his books and treasures, honoured and be-
loved by all.   In 1748, he was honoured by a visit from
the prince and princess of Wales, the father and mother of George III. ; and a glowing account of the treasures, both of nature and art, which he displayed to his
illustrious guests, may be found in a letter from Dr.
Mortimer, secretary to the Royal Society, in the *Gentleman's Magazine* for July, 1748.   In his comparative
retirement at Chelsea, and the tranquil occupation which
suited his years, he attained the advanced age of ninety-
three ; his death having taken place on the 11th of
January, 1753.   He was buried at Chelsea, in the same
vault with his wife, whose decease had preceded his by
about twenty-nine years.   They left two daughters, the
eldest of whom, Sarah, was married to George Stanley,
esq., of Poulton, in Hampshire; and the younger, named
Elizabeth, to lord Cadogan.   By his last will, sir Hans
bequeathed the whole of his museum to the nation, on
condition that 20,000*l.* should be advanced for the use
of his family, who might be considered to be deprived
by such a gift of a large portion of their patrimony.
He himself states, that the first cost had not been under
50,000*l.*   An act was immediately passed, authorising
the purchase; and in order to take charge of this collection, as well as some others obtained nearly at the same
time, several individuals were incorporated by the name
of " Trustees for the British Museum."—J. D.

SMEATHMAN, HENRY.—*Entomology.*

Celebrated as being the first to make known the extraordinary history of the *Termes*, or white ants; which,
from having studied them in Western Africa, he has
done most completely.   His paper, originally published
in the Phil. Trans. (vol. lxxi.), was subsequently translated into French, as a separate 8vo. volume, by Dr. Ri-
gaud, with copies of all the plates.   Smeathman was
one of those collecting naturalists sent abroad by Drury.
He was some time in Western Africa, and sent home a

number of superb insects; many of which were published in his patron's *Illustrations.*

Smith, Hamilton. — *General Zoology.*

A lieutenant-colonel in the British army, no less distinguished for his knowledge of quadrupeds, than for his immense collection of drawings made by himself in all parts of the world. His acquaintance with the ruminating animals is far greater than that of any living naturalist who has hitherto written upon them. It is much to be regretted that this, the only dissertation he has yet published, is incorporated in Griffith's translation of Cuvier, a work so undigested in all its other portions. He is justly characterised by Cuvier — " très savant naturaliste."

Soldani, Ambrose. — *Conchology.*

Professor at Sienna; one of the few naturalists who have investigated microscopic shells.

1. *Saggio Orittografico* overo Osservationi sopra le Terre Nautilitiche, &c. Sienna, 1780. 4to.

2. *Testaceographia* ac Zoophytographia parva et Microscopica. Sienna, 1789—1798. 3 vols. small folio.

Sonnerat. — *Travelling Collector.*

1. *Voyage à la Nouvelle Guinée.* Paris, 1776. 1 vol. 4to. With 120 plates.

2. *Voyage aux Indes* Orientale, et à la Chine. Paris, 1782. 2 vols. 4to. With 140 plates. These works, although often cited by the French authors, are very poor; the descriptions vague, and the figures, particularly of the birds, below mediocrity.

Sonnini, C. S. — *General Zoology.*

An engineer officer of the French army, and a very able zoologist; died in 1814.

Voyage dans la Haute et Basse Egypte. 3 vols.
8vo. Paris, 1799. With an atlas of 40 plates,
many of natural history.

His edition of *Buffon*, particularly the ornithological
portion, is one of the best that has ever been published.

SOWERBY, JAMES. — *General Naturalist and Artist.*

An eminent botanical painter, and a zoological draftsman ; whose plates (mostly etched by himself) have
illustrated every branch of the natural history of this
country with truth and accuracy. Mr. Sowerby joined
the most persevering industry with a surprising rapidity
of execution ; and by his professional exertions, acquired
a very respectable property. He died in 1824, and has
been succeeded in his business by one or two of his
sons.

　　The British Miscellany, or coloured Figures of
new, rare, or little known Animal Subjects. 1 vol.
8vo. London, 1806. Mr. Sowerby also executed the
plates to Wood's *General Conchology,* of which only
one volume was published. His botanical works were
numerous.

SOWERBY, G. B., FATHER AND SON. — *Conchology.*

Bookseller and commercial naturalist, settled in London ; son of the last, and one of our most acute conchologists. He has illustrated the conchological system of
Lamarck in a very useful and popular manner ; and has
not only suggested many judicious improvements, but
has defined several new genera with judgment and precision.

　　1. *The Genera of Recent and Fossil Shells.* London,
1822. 8vo. This work appears in monthly numbers,
each containing five plates, with corresponding descriptions.

2. *Catalogue of the Tankerville Collection* of Shells, with an Appendix, containing the Descriptions of several new Species. London, 1825. col. pl. 8vo. Mr. Sowerby has likewise commenced a *General Conchology*, but on the judicious plan of making each number a distinct monograph. Of these, two parts have been published; one upon the genus *Ancillaria*, the other on the typical volutes. He is likewise the author of numerous papers in the Zoological Journal, Proceedings, &c.; and has assisted his son in the following useful work:—

3. *The Conchological Manual*, by G. B. Sowerby, Jun., illustrated by upwards of 500 Figures. London, 1839. 1 vol. 8vo. The figures, without being very highly finished, are remarkably characteristic. We hope, in a new edition, the author will be more explicit upon the genera *not* proposed by Lamarck, Sowerby, Leach, &c.; and simplify his work by discarding many artificial divisions, taken from certain French authors, which have been neither followed nor heeded in this country. Young Mr. Sowerby has also published some *Illustrations of Conchology*, in numbers, which we do not possess, and have therefore not been able to quote.

SPALLANZANI, L.

A celebrated naturalist of the last century, noted for his experiments upon animals and vegetables, and his numerous writings on other subjects. He was successively Professor at Reggio in Calabria, at Modena, and finally at Paria. At this latter place, his moral character received a stain; having been detected in removing several valuable articles from the public museum. He was born in 1729, and died in 1799.

Opuscoli di Fisica, Animale e Vegetabile. 1776. 3 vols. 8vo. There is a French translation of this work by Sennebier, Geneva, 1787.; and another in English, London, 1784.

SPARRMAN, ANDRE.— *Ornithology, &c.*

Andrew Sparrman, one of Linnæus's travelling pupils, was a native of Sweden, born in the province of Upland, about the year 1747. He studied medicine at Upsal, where he acquired a great love for natural history, inspired by the prelections of the presiding genius of the place. To gratify this taste, he eagerly availed himself of an opportunity of making a voyage to China, in a vessel commanded by a relation of the name Ekeberg. So far from satisfying, the voyage only tended to increase, his desire to travel; but the want of sufficient means kept him for a time in Sweden, till an opportunity occurred of visiting the Cape of Good Hope, in the capacity of tutor to a family resident there. He arrived in the end of April, 1772; and soon after had the gratification of meeting his countryman Thunberg, in whose company he made several exploratory excursions in the neighbourhood of Cape Town. When Capt. Cook touched at the Cape, in his second voyage, Sparrman was visited by the two Forsters — father and son, — who had little difficulty in persuading him to accompany them, by the offer of a free passage, and a share of all the natural history collections that should be made. It was thus that he had the gratification of circumnavigating the globe. He was landed at the Cape on the return of the expedition, and began to practise medicine: but this was done principally with the view of obtaining the means for further travel; which he had no sooner realised, than he set out for the interior of Africa, and penetrated to 20° 30″ S. lat., about 350 leagues north-east from the Cape. This expedition was very fruitful in new plants and animals. Towards the close of 1775, he returned to Sweden, where he obtained the degree of M.D., and was elected a member of the Royal Society of Stockholm, of which institution he was afterwards appointed president. The charge of baron de Geer's collection of natural history, which had been bequeathed to that society, was likewise confided to him. Some years before his death, he joined an expedition designed to explore Western Africa; but it led to no important result. He died at Stockholm, on the 20th of July, 1820.

Museum Carlsonianum; novas et selectas Aves exhibens. Holmiæ, 1786—89. small folio. The figures, in general, are good; but the descriptions are too concise to be of much value to the modern ornithologist.

Spence, William.—Entomology.

The well-known coadjutor with Mr. Kirby in the " Introduction to Entomology," published in their joint names. Mr. Spence has also written an admirable monograph, in the Linnæan Transactions, on the genus Choleva.

Spinola, Maximilian. — Entomology.

A descendant of the noble and illustrious Genoese family of that name. An amiable man, and a profound entomologist. He is the possessor of a most extensive library, and of a rich collection of the hymenopterous insects of Italy.

Insectorum Liguriæ, Species novæ aut rariores. Genuæ, 1806—1808. 2 vols. 4to. fig.

Spix, J. B. de.

An able naturalist and indefatigable traveller; who, with Dr. Martius, the botanist, was sent by the Bavarian government to explore the productions of Brazil, where they travelled for several years. They returned to Europe with immense collections, and soon began to publish their discoveries; but the death of Dr. Spix rendered it necessary to call in other assistance. Wagler undertook the reptiles, Agassiz the fishes, and Perty the insects. All these volumes are published separately; but the execution of the plates bears no comparison to the extravagant price of all these works. As matter of curiosity we have annexed these prices, as given in Bohn's Catalogue.

1. *Semiarum* et Vespertilionum Braziliensium Species novæ (Latin and French). 1 vol. royal folio. 38 plates. Monachii, 1824. 10*l.*

2. *Avium* Species novæ, quas in Itinere Annis 1817—20, per Brazilium collegit et descripsit. 2 vols. royal 4to. 222 coloured plates. 1824—26. The figures are not above mediocrity, although highly valuable for consultation; and the reduced price (42*l.*) is enormous.

3. *Serpentum* Braziliensium Species novæ. Curante J. Wagler. Monachii, 1824. 28 plates. 7*l.* 15*s.*

4. *Testudinium* et Ranarum Braziliensium, Species novæ. 1 vol. Monachii, 1824. 39 plates. 7*l.* 10*s.*

5. *Animalia* nova sive Species, novæ Lacertarum in Itinere per Brasilium collecta. 30 plates. Monachii, 1825. 6*l.* 6*s.*

6. *Testacea* Fluviatalia. Digessit, descripsit et Observationibus illustravit J. Wagler. Ediderunt F. Schrank et C. F. P. Von Martius. 29 plates. Monachii, 1827. 4*l.* 4*s.*

7. *Piscium* Brasiliensium. Selecta Genera et Species. Digessit, &c. L. Agassiz. Monachii, 1829—31. 2 vols. 101 plates. 18*l.*

8. *Delectus Animalium* Articulatorum. Digessit, &c. M. Perty. 3 parts, 36 plates. 12*l.* 12*s.* Making altogether 10 volumes, royal 4to., the *reduced* price of which is 108*l.* 7*s.* How can it possibly be supposed that such publications advance science by diffusing knowledge, when their price amounts to a prohibition to all but wealthy amateurs, and renders their consultation all but impracticable?

SPRY, WILLIAM. — *Flower Painter.*

One of the most promising botanical and entomological artists of Britain. His outline figures to Shuckard's work on *British Coleoptera,* are not only remarkably accurate, but executed in a style peculiar to himself. They are perfectly sufficient for all purposes of science ; and are so cheap, that they deserve being taken as models for delineating this order of insects.

STEDMAN, J. G.

A captain in the Dutch service, who wrote an interesting account of his campaign in the forests of Surinam. Although no naturalist, many interesting observations on the native animals are scattered through his work.

Narrative of a Five Years' Expedition against the revolted Negroes of Surinam, from 1772—1777, including a History of that Country, and describing the Productions. London, 1796. 2 vols. 4to. Another edition was published in 1806. It is full of plates, but they are poorly executed ; and those of the animals, &c. are very bad.

STEPHENS, J. F. — *Entomology.*

A laborious and very zealous entomologist, possessing a very large collection of British insects.

1. *Systematic Catalogue* of British Insects. 1 vol. 8vo. London, 1829. The most complete list, we believe, hitherto published of British insects, but the synonyms, in many instances, cannot be depended upon, having been compiled from other authors.

2. *Nomenclature* of British Insects. 12mo. London, 1829.

Ditto, second edition, Part I. 8vo. London, 1833.

3. *Illustrations* of British Entomology: — Mandi-
bulata.   6 vols., and part of vol. 7., 8vo.   1828—35.
— Haustellata. 4 vols. 8vo. 1828—35.

Mr. Stephens has also compiled the concluding ornithology of Shaw's General Zoology.

STOLL, CASPER. — *Entomology*.

A medical practitioner at Amsterdam, and a zealous
entomologist.

1. *Representation*, exactement coloriée d'après
Nature, des Spectres, des Mantes, des Sauterelles, &c.
Amsterdam, 1787.   8 parts, forming 1 vol. 4to.

2. *Representation*, exactement coloriée d'après
Nature, des Cigales et des Puniases, &c. Amst. 1780.
12 parts, forming 1 vol. 4to.

Both these works are somewhat scarce, particularly
the former.   The figures are not well drawn ; but
they are valuable, as being uniformly quoted by Fabricius, and as few are contained in other works.
Stoll likewise edited the 5th or supplementary volume
to Cramer's *Exotic Insects*, which has much greater
merit, and is of less frequent occurrence than the
original work.

STORR, G. C. C. — *Mammalogy*.

Prodromus Methodi Mammalium. Tubingæ, 1780.
4to. pp. 43. pl. 4.

STRAUS-DUERCKHEIM, H. — *Entomology*.

An eminent and philosophic writer, remarkable for
his bold deductions and able generalisations.

Considérations générales sur l'Anatomie comparée
des Animaux Articulés, aux quelles on a joint l'Ana-

z

tomie descriptive du hanneton vulgare (with an atlas of 10 plates). 1 vol. 4to. Paris, 1828. The plates are beautifully executed, and the whole is said to be the most complete and accurate entomographical work yet published.

STURM, JACQUES. — *Entomology.*

An excellent entomological artist, and an accurate observer of insects.

1. *Deutschlands Fauna.* Nuremberg, 1807. 4 vols. 8vo. pl. col. 52.

2. *Insecten Sammlung,* &c. Nuremb. 1800. 1 vol. 8vo. pl. col.

SULZER, J. H. — *Entomology.*

Du Kennzeichen der Insecten, &c.; or, The Characters of Insects (in German). Zurich, 1760. 4to. fig.

SWAINSON, WILLIAM. — *General Zoology.*

I had intended the portrait of the illustrious count Maurice of Nassau to have been the vignette for this volume; but as the publishers, in their partiality, have requested my own, I have no alternative than to say something about myself. Autobiography, in its very nature, cannot be otherwise than egotistical : and yet, it is perhaps the most authentic of all records; for, as history is based upon facts, who so likely to give these correctly, as the individual to whom they relate? * With inferences he has little to do, and with opinions still

---

* As a further apology for this notice, the reader may perceive in the *Règne Animal,* that I am stated to be the author of several papers in the Linnæan Transactions (where not one of mine is to be found); and that I wrote, in conjunction with Dr. Horsfield, a memoir on the birds of Australia, a remark which applies to Mr. Vigors—not to myself.

less: these belong to the public, who generally decide such matters for themselves. I was born on the 8th of October, 1789. My mother's name was Stanway; she died in the flower of life; — a tall, elegant, and beautiful creature, taken from this world ere she was twenty-five, and before I knew her loss. My father's ancestors, time out of mind, had lived on their lands near Hawkshead, in Westmoreland; but by the improvidence, or misfortunes, of one or more, the paternal estates gradually passed into other hands. He, as well as my grandfather, filled various posts in the Custom-house; and the former only resigned the collectorship of Liverpool,—the most important in the gift of the crown, — two or three years before his death. I was destined for the same line of service, and was appointed junior clerk in the secretary's office (then held by my father), with a salary of 80*l.* a year, at the early age of fourteen: my education, in fact, from unavoidable circumstances, was left unfinished. An impediment of speech, resulting from a peculiarly nervous temperament, acted as an insuperable bar to the acquisition of languages, and I showed not the least aptitude for the ordinary acquirements of schools. Hence it was, that at an age when other young men of my own standing were at college, I had entered public life. But I was wayward and unhappy. With prospects of rapid advancement which might well be envied, I had not the least inclination to pursue them. My father had a collection of British insects and shells, and these had given me not merely a taste, but a passion, for natural history even when a mere child; and every moment I could command was divided between drawing and collecting. It was in vain that my parents endeavoured to repress this ardour, and to make these tastes subordinate: their judicious restraints only increased the evil: sleeping or waking, my thoughts were constantly bent on how I could get abroad, and revel in the zoology of the tropics.

About this time, I remember to have read Smeath-

man's notes on the insects of Western Africa.* These
so excited me, that I copied them out, and thought no
earthly happiness could be greater than visiting Sierra
Leone, and capturing thousands of butterflies; or going
out to some distant country, even to collect for others.
With such wild and uncontrolled ideas, it cannot be supposed that my official duties were performed as they should
have been. My father saw this; and as his friend, commissary-general Wood, was about proceeding to join the
Mediterranean army, he got me placed on that establishment by a Treasury minute, and my situation in the
customs was at once resigned. It was now only that
happiness seemed to be before me. I knew that I
must do my duty; but then, in all other respects, I should
be my own master. Filled with these anticipations, I
sailed for the Mediterranean, in the suite of the commissary-general, and landed in Malta, from whence, after
a short stay, we at once proceeded to Sicily, in the spring
of 1807.

The British army then merely garrisoned that island,
without undertaking any very decided operations against
the French, who were in possession of all Calabria.
Hence our duties were comparatively light; we lived in
comfortable quarters, and enjoyed much leisure: this
continued, with very little intermission, for several years,
during which I alternately investigated the zoology and
botany of that charming island. An annual leave of
absence of six weeks or two months enabled me to visit
Greece,—the botany of which classic region had been
rendered more interesting from the appearance, about
this time, of Dr. Sibthorp's *Prodromus*. My expectations of Sicily, as a field for zoological research, had
been somewhat disappointed: it is a perfectly wood-
less country, and almost destitute of permanent rivers:
but the beautiful little streams which meander through
the Peloponnesus, have their banks constantly moist and
verdant; hence plants and insects abound. Our journey,
nevertheless, was a hurried one; and although my bo-

* In Drury's Illustrations of Entomology.

tanical acquisitions remain, the beautiful insects I collected, ultimately fell a prey to *Dermestes*. I had
scarcely returned to Malta, before the plague broke out
in the capital. The quarter in which I resided, was
one of the most infected ; the street was barricaded,
and for near two months I was a complete prisoner.
Provisions were brought in by the authorities, and received into the house by an opening cut through the door.
At last it became so destructive, that the cart which
conveyed the dead away, came round to be filled every
day ; and it was no uncommon spectacle, upon rising in
the morning, to see half a dozen dead bodies laid on the
pavement, on both sides of my own house, ready to be
removed. I know not how it was, but I felt more dismay on the first death by this scourge, than by the subsequent horrors of such fearful sights. Confined to the
house, with only one domestic, I substituted, for my
usual daily exercise of walking and riding, that of carrying some loose stones left in the yard by the masons,
from thence to the top of the house, and then down
again. I thought seriously ; placed my trust in that
Providence which had hitherto preserved me ; and
felt not only resigned, but perfectly tranquil, to whatever
might happen. This imprisonment enabled me to finish
many of my Sicilian and Grecian sketches, and arrange
the plants and animals. In short, I was almost sorry,
on my own account, when our street was released from
quarantine, and I had again resumed my official duties.
The withdrawal of the French from Italy, by the
united operations of our troops with those of Austria,
required my services with the army in Naples, and I
had thus an opportunity of treading the soil of Italy.
Soon after this, I was appointed, by general Maitland,
chief of the commissariat staff in Genoa. But the
glorious works of the Italian painters, so profusely scattered in the churches and galleries of Rome and Florence, cooled, for a time, my passion for natural history.
I began collecting their pictures, sketches, and etchings,
— particularly those of the Genoese school, — without,

however, neglecting the plants and insects of northern
Italy. After making excursions, as opportunity or
duty permitted, through various parts of Tuscany, I
was again ordered to join the head-quarters of our army
at Palermo, where I arrived in the autumn of 1814.
The Russian campaign of 1812 had now totally changed
the political horizon. England had at length restored
Naples to the king of the Two Sicilies ; and the French
had been completely driven out of Italy. Eight years
had elapsed since I quitted England; and I looked forward, with no small delight, to the reduction of the
Mediterranean army, which would release me from my
official duties. The examination and audit, however, of
the numerous accounts connected with its establishment,
required the greatest exertion on our part; and my application to return home was therefore suspended until the
following year. At Palermo, I had the pleasure of
meeting the baron Bivona, the most learned botanist of
Sicily; and my old correspondent, Rafinesque Schmaltz,
whose first name is familiar to most zoologists. In the
society of such congenial minds, I passed many happy
hours, and made many delightful excursions. By the
assistance of the first, my materials for a *Flora Sicula*
were considerably augmented ; while, by the induce-
ments of the latter, I was led to investigate the ichthyology of the western coast. These duties and relax-
ations continued until the middle of 1815; when my
health gradually getting worse, it was deemed necessary,
by the medical men, that I should return to England.
I embarked from Palermo; and had the happiness of
landing all my collections of nature and art at Liverpool,
in the autumn of 1815.

I was now only twenty-six; and through the powerful
interest of my family connections (certainly not from my
own merit), I had risen to a rank somewhat unusual
for so young a man.* I liked the service, but my old

---

* I was, in fact, the youngest Assistant Commissary-General on the staff
of the Mediterranean army. I may be pardoned, therefore, for having been
somewhat particular in my horses and " equipments."

passion for travelling in tropical countries returned with
its original force : I had now to choose, whether I
would give up the latter for some new and higher appointment my friends were ready to procure me, or
whether I should go upon half pay, and follow my own
course. I hesitated not to choose the latter. After
living so long upon the Continent, and accustomed to
the unsettled life of a soldier, I was struck by what I
thought the artificial habits and the luxury of English
society. I sighed for my Sicilian cottage ; I longed again
to ramble over mountains clothed with luxurious plants
—to sketch delightful scenery—to rise with the sun,
gallop on the sands, climb precipices, and swim in the
sea. In place of this, I had to join dinner parties,
drink wines I detested, ride in carriages, dance at balls,
and do a hundred other things for which I had neither health nor inclination. Domestic society I truly
enjoyed ; but that was not sufficient to keep me at
home. I had, therefore, no sooner returned to England,
than I began laying plans for quitting it. Having
been delighted with reading Le Vaillant's travels, when
a boy, and subsequently perusing those of Mr. Barrow,
I fixed upon Southern Africa as the best field for zoological investigation. I therefore began reading books,
and filled a volume with extracts of every thing about
the Cape. But this project was diverted by a singular
incident. Happening to spend an evening with Mr.
Lambert, the celebrated botanist, he told me he had just
had a letter from a friend of his, who had been many
years travelling at the Cape, and had brought with him
a collection which filled two waggons ! This friend was
no other than Dr. Burchell. I heard the news with
dismay ; for what, thought I, can be now left in South
Africa, more than the gleanings of a harvest already
reaped ? A little consideration might have showed me
the absurdity of this opinion ; but as I could not submit
to follow in the wake of another, I at once determined
to relinquish the Cape, and choose some other quarter
yet untrod by the naturalist. This choice was soon

made. About this time, the jealousy of the Portuguese government relaxed, and they opened Brazil to European researches. Mr. Koster had just published his travels : he gave me such a picture of the zoological riches of the country he had just quitted, that I resolved to accompany him on his second journey; and we left England together on the 22d of November, 1816. To give all the particulars of my subsequent travels would be tedious. Suffice it to say, we landed at Pernambuco ; where Mr. Koster had no sooner purchased a small plantation, than the memorable revolution of 1817 broke out. As the peace of the whole province was thus disturbed, he deemed it prudent to give up his original intention of travelling across the Continent. The English, indeed, were respected, and had nothing to complain of ; but still travelling in the interior was dangerous. Meantime, and until this outbreak against a corrupt and wicked government was quelled, I remained in the vicinity of Olinda, finding ample occupation in collecting plants and animals. The insurrection being put down, I immediately engaged a guide and three Indians, with whom I set off, overland, for the Rio St. Francisco. We found the draught, however, so great, that we were obliged to reach Bahia by water.* After investigating several parts of that province, we proceeded by sea to Rio de Janeiro. Here I met with Dr. Langsdorff, the late Dr. Raddi of Florence, and some of the German naturalists sent by the court of Austria. With Langsdorff I made several excursions, and in four months so enriched my collections, that I became almost satiated. I felt I had now more than enough to study and arrange for years to come. I therefore broke up my party, embarked for England, and once more,—like a bee loaded with honey —returned to my father's house.

The multiplicity of my collections made me uncertain

---

* While travelling the *Sertem*, or interior of Pernambuco, we were constrained to drink what would be termed in England *ditch* water ; hundreds of cattle perished ; whole villages migrated to the sea coast ; and we often were obliged to pick the maggots out of our dried meat before it could be converted into soup.

what to do first.   I sent a short abstract of my travels
to Professor Jameson, at his own request * ; but, as he
printed it without any comment, or one word of praise, I
abandoned all intention of publishing them in a separate
work.   I was discouraged by the idea, that the unpatronised researches of an unknown individual might probably be thought insignificant, when compared to those
of naturalists sent out by governments, and which the
editor lavishly praised in the very same number of his
Journal.†   I mention this, to show how the feelings of
young authors may be influenced, and their energies repressed, even by such indirect discouragements.   This,
in fact, was the true reason why my narrative was never
published.   Soon after my return, and at the recommendation of sir Joseph Banks, I was elected a F.R.S.,
having been admitted a fellow of the Linnæan before
I embarked for Brazil.

It was about this time that the art of lithography was
first introduced into England. Encouraged by my friend
Dr. Leach, I determined to try how far it might be
used in producing zoological plates fit for colouring.
My attempts succeeded ; and the first series of the *Zoological Illustrations* was the result.   As I took upon
myself the whole expenses and management of this work,
I soon found that its publication, in monthly numbers,
rendered it necessary I should superintend all its mechanical details : I therefore quitted Liverpool, and took
lodgings in Surry Street, Strand, where I lived nearly
the life of a " hermit in London" for two or three
years.   The late hours, and style of visiting customary
in the metropolis, neither suited my health, or the steady
prosecution of my work.   I laboured hard, during the
greater part of the year, to enjoy the leisure of autumn
among my family and friends.   My little book was favourably received ; and, thus encouraged, I brought out

* Edinburgh Philosophical Journal.
† Never, perhaps, was so little done by such a party. Out of five or six
naturalists, sent from Vienna to investigate the botany and zoology, the
only one who remained sufficiently long to reap the harvest before him,
was my friend, M. Natterer ; the rest, after wasting their time at Rio, and
making little excursions in the province, returned to Europe.

the early numbers of *Exotic Conchology*. The uncertainty, however, which then attended the lithographic process was so great, that after being frequently obliged to draw the same subject two or three times before the printer produced a tolerable impression, I was compelled to suspend the publication, and confine myself to the *Illustrations*.

Before my removal from Liverpool I had formed an attachment to the only daughter of John Parkes, Esq., of Warwick; but we were not exempt from those difficulties which so frequently impede marriage, particularly in this country. A vacancy in the British Museum, about this time, caused by the deplorable illness of my friend Leach, induced me to apply for the appointment. I produced the highest testimonials from such men as Cuvier, Roscoe, Dr. Rees, sir James Smith, Dr. Trail, sir W. J. Hooker, Dr. Scoresby, and numerous others. But I was refused, and a gentleman (I. G. Children, Esq. *), who knew nothing of natural history, was appointed chief of the zoological department. As a faithful historian I am bound to mention this fact, without the least unkindly feeling against that individual. The disappointment, indeed, at the time, was acute; but I have lived to rejoice it was so ordained by Him, who foresees consequences we have no conception of. Frustrated in this hope of adding to a small independence, I determined no longer to wear out the rest of youth in longing for domestic life : my gentle friend thought the same, and we were married in the autumn of 1825. My venerable father expired the next year, and as my portion, although the eldest, fell very short of what we had expected, we found it prudent, for a time, to avoid the expense of a separate establishment. The annual allowance of 200*l.* from my father having

---

* On the retirement of this gentleman a few months ago, I again applied for the situation, not from the remotest idea of retaining it, but that by holding it for six months previous to my final departure from England, I might submit to the trustees a total change in the management of the zoological department, and then resign. I was not, however, so far honoured as to receive any notice to my application.

suddenly ceased, I now began to think seriously for the future. Bred up with somewhat of aristocratic notions, and accustomed, when on service, to *command* rather than to *obey*, I had a rooted dislike to all commercial affairs, and would rather have gone once more on active duty than have sat behind a desk. At length, it occurred to me that no profession was more honourable than that of an author ; that many of my friends found it a source of profit, no less than of fame ; and that I might justly turn to pecuniary account that knowledge, to gain which I had sacrificed so much. One of my friends, accordingly, took an opportunity of mentioning my views to the house of Longman, Orme, Brown, and Co., the first publishers and booksellers in the kingdom ; and this led to a connection which has continued to the present moment. Hitherto I had written for amusement, I was now to write as a professional author. I felt so diffident of my powers in this new walk, that, after revising the entomological portion of Loudon's two *Encyclopædias of Agriculture* and *Gardening*, I absolutely declined proceeding further, from a sense of incapacity ; and here the whole business would have terminated but for the encouragement of Mr. Longman, who expressed that satisfaction with my performance which I certainly did not entertain myself. An *Encyclopædia of Zoology* was next proposed, to match with those of Loudon's, for which I was to execute all the drawings upon wood: it was an Herculean task, but I undertook it. On this work I laboured incessantly for several years, and had brought it nearly to an end, when an unexpected event stopped its progress, or rather made it assume a totally different form.

The editor of the Cabinet Cyclopædia had resolved to engage a party of naturalists, to execute the zoological series ; and a long list of names had been given him, many of the highest repute, who were to take the several portions. The unfriendly feeling entertained towards me by an individual, whom the editor empowered to organise this undertaking, was (I am well assured) the reason why my assistance was never asked. Promises, indeed, were continually made to the editor that some of these volumes would be " soon " ready; but after waiting near three years, not even one was forthcoming. Finding, therefore, that, from some unexplained cause, nothing was produced from this imposing array of great names, recourse was had to the only naturalist of the least repute, whose services had not been thought worth securing. To make " a long story short," a proposition was made to me, that the *Encyclopædia of Zoology* should be remodelled, and transformed into the *Cabinet of Natural History.* I foresaw that this would almost impose on me the necessity of re-writing the whole work; but I felt flattered in being thought equal to the task, and in having the whole series under my controul,—after having been shut out from the least participation in it: the agreement was therefore concluded, and the public are in possession of the result.

The different aspect, however, under which my labours were now to appear, soon convinced me I was called upon to do more than was " in the bond." It was highly desirable that a uniform system of arrangement should pervade the whole series. I had thus only the choice of following the *Règne Animal,* or of working out, as far as possible, that system I had already adopted in theory, and partly exemplified in detail; namely, the circular arrangement of animals on the principles of their affinities and analogies. I hesitated not to attempt the latter, and here began the most arduous period of study of my whole life.

I had quitted Warwick and settled at Tittenhanger Green*, a spot so retired, as to be completely out of the reach of morning visitors. Here, surrounded with immense collections and a large library, I had all the materials of study under my own roof: my facilities were great, and I improved them by occasionally visiting the collections in London. I employed near six years in working out my theory, elsewhere explained, through

* Within a mile of the little village of London Colney, Herts.

all the different classes of animals; and thus prepared, I ventured to give the result to the public in the *Preliminary Discourse.* I mention this, not from attaching any great merit to the thing, but simply to account for what may appear an undue degree of confidence in the opinions therein expressed, before the reader had been put in possession of the facts upon which the whole theory was founded. I verily believe, that, had I expressed my convictions in a more subdued tone, many of those who now differ from me would have adopted these views, — at least in a general way; but I am always so delighted with detecting either a new link of relation, or in bringing an isolated fact to bear upon general principles, that my enthusiasm sometimes overcomes my judgment. I forget, in fact, that no one, unacquainted with the other instances of a similar nature, — all converging to the same point, — can possibly attach the same importance to a *single* instance, that I do myself.

While slowly elaborating this new disposition of the animal world, and regularly proceeding in my chief undertaking, I yet found time to attend to others, at short intervals of leisure. Among these were the zoological portion of Murray's *Encyclopædia of Geography,* the *Birds* of the "Northern Zoology," and the second series of the *Zoological Illustrations;* to these may be added the two volumes on the *Birds of Western Africa,* and another on the natural arrangement of the *Flycatchers,* published in Lizars's "Naturalist's Cabinet." In 1828, I was induced to spend six weeks in Paris, for the purpose of studying the insessorial birds contained in that superb collection. By spending seven hours daily in the Garden of Plants, I succeeded in making drawings and descriptions of nearly every species I did not possess; and thus fortified, I ventured to give the outlines of my views on their natural arrangement in the *Northern Zoology.* I cannot omit mentioning in this place, the excessive liberality I experienced from Cuvier, Geoffroy St. Hilaire, and all those eminent men attached to the museum, who had the power of facilitating my

researches, or of making my short stay in Paris agreeable. Dr. Isidore Geoffroy, in particular, gave up to
me his own little study in the museum, in which I was
permitted to remove every specimen from the gallery I
desired to examine, with as much freedom as if they
had been my own.

Let it not be supposed that the retired life which these
pursuits have obliged me to adopt for the last fifteen years,
has been a period of uninterrupted tranquillity; or that
by withdrawing, almost, from the world, I have escaped
its vexations and trials. Far from it. The gradual loss
of nearly half my fortune by the utter failure of two of
the Mexican mining companies, once the most promising*, would hardly deserve mention, save to warn others
against faith in the names and promises of joint stock companies; and to record that this loss has been recompensed
by more fortunate investments. So true it is that an Al-
mighty Providence makes all things to work for good,
— "to those who love God." Far, very far, greater was
that trial, laid upon me in 1835, when I became a
widower with five children. No husband *could* have
been happier during twelve years. But He, who takes
away, can console. It is to watch over these living testi-
monies of our love, to preserve them in those simple
habits and affectionate feelings, which alone constitute
true enjoyment — to teach them from experience,
that the paths of virtue, founded upon religion, are alone
those of happiness, — it is to accomplish *such* objects
that I am about to transplant myself and them to a new
soil, in the southern hemisphere. Should this be carried into execution, the parent trunk will there fall;
yet it will be surrounded by scions who may perpetuate
its name and lineage.

The greater part of my collections, I trust, will be
transported to New Zealand, where they may possibly

* That of Real del Monte and Bolanos — ruined by the bad management
of the "manager" and his supporters in England. That the proprietors
can consent to sink more of their money, under such superintendence,
would be incredible — were it not true.

stimulate others to the study of nature, and form the
basis of a Zoological Institution.   My career, as a professional author, will soon close.* The motto, prefixed
to this volume, conveys the result of my experience.
The measure of talents, whether small or great, with
which a man is intrusted, is but "vanity and vexation
of spirit," unless employed to the honour of that Being
who has bestowed the gift.   Nor can the highest fame,
or the greatest prosperity, counterbalance that internal
peace which this conviction will alone produce.

1. *Instructions* for collecting and preserving Subjects of Natural History and Botany. Liverpool, 1808.
Privately printed.

2. *Zoological Illustrations*, or Figures of new, rare,
or remarkable Animals. London, 1820—23. 3 vols.
royal 8vo.

3. *Exotic Conchology*, or coloured Lithographic
Drawings of Shells. London, 1822—35. Royal 4to.
complete in 6 parts.

4. *The Naturalist's Guide* for collecting and preserving Subjects of Natural History. London, 1824.
12mo.

5. *Zoological Illustrations*, Second Series, complete
in 36 Nos. or 3 vols. royal 8vo. London.

6. *Ornithological Drawings*; the Birds of Brazil.
London, 1834. 5 parts, royal 8vo. The 6th, which
terminates the series, is almost ready.

7. *The Geographical Distribution* of Man and of
Animals, in Murray's Encyclopædia of Geography.

8. *Fauna Boreali-Americana*, or Northern Zoology. Part 2. the Birds. (The plates and the greater

* That my foreign correspondents may not construe this into a *total*
abandonment of zoological pursuits, I still hope to communicate with them
as heretofore; I shall be most happy to exchange duplicate insects, &c.,
particularly from India, America, the Cape, and different parts of Australia. For this purpose, parcels sent for me to England, should be directed to
the care of Messrs. Longman, Orme, and Co., Paternoster Row; or, to W.
Shuckard, Esq., librarian to the Royal Society, Somerset House, London.
But as Sydney is the most direct channel of communication between New
Zealand, India, and the Brazils, any thing sent from those quarters may be
addressed to the care of Mr. Reid, chemist, Sydney; or to the "care of the
Officer in charge of the Commissariat of Accounts, Sydney, New South
Wales."

part of the letterpress : the specific descriptions being
by Dr. Richardson.) 1 vol. 4to.

9. *Observations on the Natural System,* being the
Introduction (printed separately), to the above volume.

10. *Elements of Conchology,* for the Use of Students
and Travellers. 1 vol. 12mo. London, 1834.

11. *Birds of Western Africa.* 2 vols. 12mo. with
coloured plates. Edinburgh, 1837.

12. *Flycatchers,* the Natural Arrangement and
History of. 1 vol. 12mo. coloured plates. Edinburgh,
1838.

13. The whole of the Volumes of the *Cabinet Cyclopædia* of Natural History hitherto published, of
which this is the eleventh.

Several papers of mine will be found in scientific
Transactions, Journals, and periodicals, the names of
which I cannot now remember. The most useful, perhaps, is the following : —

Synopsis of the Birds of Mexico, brought to
England by Mr. Bullock, in Taylor's *Philosophical
Magazine,* No. 15. for June, 1827. This paper was
published long before any of these birds reached the
Continent, where they have been described by Wagler
and others, under different names.*

SWAMMERDAM, JOHN. — *Entomology.*

This distinguished anatomist and physiologist, who
was among the first scientific men who applied the
microscope to the examination of the minuter parts of
animal structure, and whose consummate skill and
indefatigable perseverance effected many important discoveries, was the son of John James Swammerdam and
Barentje Corver; and born at Amsterdam, on the 12th

---

* It is not to be supposed that this is a *wilful* oversight; but it illustrates
the impracticability, which is daily increasing, of ascertaining what subjects
are really new; and where, amid countless publications, an author can discover that which he is in search of, and which he feels bound to consult.

of February, 1637. His grandfather obtained the sur-
name of Swammerdam from the place of his birth, a
village on the Rhine; and it continued to be applied to
his descendants ever afterwards. His father was an
apothecary; and having acquired some fortune, expended
a portion of it in collecting a museum of natural history,
and other objects of curiosity. Young Swammerdam
was intended for one of the learned professions; and the
church was at last chosen for him ; but being unwilling,
after due reflection, to take upon him the responsible
duties of that sacred office, he obtained his parent's
permission to study medicine. With this view, he repaired to Leyden, to enjoy the advantages of its celebrated university. Here he highly distinguished himself by his skill in anatomy, and the anxiety he displayed
in the acquisition of every kind of knowledge relating
to the physical sciences. He then visited France; residing for a time in the house of Tanaquil Faber, and
afterwards at Lyons. By the celebrated traveller, Thé-
venot, with whom he lived on terms of intimacy, he
was introduced to Van Benningen, a senator of Amsterdam, then ambassador at the court of France, who
conferred many favours on him after his return to his
native city. For some years after his return, the greater
portion of his time was devoted to the study of physic
and human anatomy ; and, in 1666, he repaired to
Leyden to take his degree, which he obtained on the
22d of February, 1667. His thesis was on Respir-
ation ; and he afterwards enlarged and published it.
From an early period of his life, he had studied the
habits and structure of insects with extraordinary assiduity; and when the grand duke of Tuscany visited
Holland, he was so much struck with Swammerdam's
skill in dissecting them, and his general attainments,
that he offered to purchase his museum, and provide
for him at his own court, if he would accompany him
to Italy. This generous offer, however, was declined ;
and Swammerdam continued his entomological investigations with redoubled ardour. In 1669, he published

A A

a *General History of Insects.* His devotion to this subject made him almost entirely neglect his professional
duties; which so displeased his father, that he withheld
from him, for a long time, any further supplies of
money. His health, also, was falling a sacrifice to his
habits of unremitting application; and was in such a
condition, as to render him almost unfit for medical
practice. In the years 1671 and 1672, his studies related chiefly to human anatomy, fishes, and insects. In
the end of 1673, he concluded his examination of the
structure of bees; in which he had laboured with such
ardour, that his health was irreparably injured. His
mind, too, was in a state of doubt and despondency
with regard to religious matters; and, while in that
condition, he adopted the opinions of Antoinette Bou-
rignon, a wild enthusiast of the day, and became one
of her most zealous disciples. He still, at intervals,
continued his studies; but he gradually became more
reluctant to engage in such pursuits; and ultimately
resolved to retire from the world altogether, and spend
the remainder of his life in solitary meditation. To
enable him to do this, he wished to sell his museum,
which was now of great value; but could find no pur-
chaser, although he again applied, through his former
friend Steno, now become bishop of Titiopolis, to the
grand duke of Tuscany. On the marriage of his sister,
his father went to reside with his son-in-law, and the unhappy Swammerdam was at length deprived of a regular
home. The death of his father afforded him the prospect of some improvement in his finances; but when the
property came to be divided, his expectations were by
no means realised. Soon after, he had a severe attack
of quartan ague, which, for a time, completely prostrated
his remaining strength; and even when somewhat recovered, he continued shut up in his chamber in a
moody hypochondriacal state of mind, during which any
allusion to the pursuits in which he formerly delighted,
excited his severest displeasure. A final attempt to
dispose of his museum having failed, he resolved to sell

it by auction : but he was not destined to witness its
dispersion ; for his constitution, which had long been
sinking, could hold out no longer, and he expired on
the 17th of February, 1680.

His manuscripts and plates he bequeathed to Théve-
not; and after passing through several hands, they were
purchased by the celebrated Boerhaave in 1727, who lost
no time in giving them to the world.   They form the
well-known work, entitled Swammerdam's Book of Nature (*Biblia Naturæ*), which the learned editor has further enriched by an interesting life of the author.—J.D.

> *Biblia   Naturæ.*   Leyd. 1737—1738.   2 vols.
> folio.   The text is in Latin and Dutch.   There have
> been various translations of this curious book.   An
> abridged edition was published at Utretcht, under the
> title of *Histoire Générale des Insectes.* 1682. 1 vol.
> 4to. pp. 215. pl. 13.

TEMMINCK, C. J. — *Ornithology.*

A celebrated ornithologist, and curator of the Royal
Museum at Leyden.

> 1. *Histoire Naturelle des Pigeons*, avec Figures
> peintes par Mademoiselle Pauline de Courcelles. Pa-
> ris, 1808. folio, pl. col. 86.   A magnificent volume.
>
> 2. *Histoire Naturelle Générale des Pigeons et des
> Gallinacés,* accompagnée avec Planches Anatomiques.
> Amsterdam, 1813—15. 3 vols. 8vo. The most com-
> plete account of these interesting birds that has yet
> appeared : the species are minutely and accurately
> described ; and many natural groups for the first time
> characterised.
>
> 3. *Manuel d'Ornithologie,* ou Tableau Systema-
> tique des Oiseaux que se trouvent en Europe.   Am-
> sterdam, 1813. 8vo. A second edition of this work,
> with very great additions and amendments, appeared
> in 1820, in 2 volumes.   To this is subjoined, an
> analysis of the general system of ornithology,

adopted by the author in this and several other publications.

4. *Observations sur la Classification* Méthodique
des Oiseaux. Amst. 1817. pp. 60. The object of
this pamphlet is more particularly to analyse the
system of M. Vieillot, whom the author accuses of
dishonesty and plagiarism towards the celebrated
Illiger. The whole pamphlet is curious.

5. *Nouveau Récueil des Planche Colorées* des
Oiseaux, pour servir de Suite en de Complément aux
Planches Enluminées de Buffon. Paris, 1820—
1823. 4to. This work appears in numbers (each
containing six coloured plates), of which are already
published. As a collection of figures, it is of much
value ; and there are occasional notices respecting the
synonyms of other species, which render its consultation very instructive : but in general the descriptions
are meagre and unsatisfactory; little notice is taken of
their natural affinities, and still less of their habits.
The plates are mostly very good, but in general too
highly coloured; and they all bear the appearance of
being drawn from stuffed specimens.

6. *Monographies de Mammalogie,* ou Descriptions
de quelques Genres de Mammifères dont les Espèces
ont été observées dans les différens Musées de l'Europe. Paris, 1825. Liv. 1—6. 4to. This is one of
the most valuable scientific works on the *Mammalia*
yet produced.

THIERY DE MENONVILLE, N. J.— *Entomology.*

A French physician, who brought the cochineal insect from Mexico.

Traité de la Culture du Nopal et de l'Education
de la Cochenille. Paris, 1787. 2 vols. 8vo. fig.

THOMAS, P.

A physician (probably of English extraction), settled
at Montpellier.

Mémoire pour servir à l'Histoire Naturelle des Sangsues. Paris, 1806. pamphlet in 8vo.

THOMPSON, DR. — *General Zoology.*

Celebrated for his discovery of the metamorphosis of certain *Crustacea*, by which that order and the *Cirripedes*, or barnacles, are proved to be connected. Dr. Thompson holds the high medical rank of Deputy Inspector of Hospitals to the troops in Australia, where it is hoped he will yet prosecute his brilliant discoveries.*

Zoological Researches. In 6 parts, forming 1 vol. 8vo.

THUNBERG, C. P. — *Zoology and Botany.*

A distinguished naturalist and traveller, and a favourite disciple of Linnæus. He was the son of a country minister, and, at an early age, evinced a strong passion for the study of nature. He visited France and Holland; and, by the pecuniary assistance of his friends, was enabled to go on his travels. He visited the Cape of Good Hope, Ceylon, Java, and Japan; and brought from those countries a rich harvest of new plants and insects. On his return, he met with that favour from his sovereign, which he had so well deserved: he was created a knight of the Order of Vasa, and Professor at Upsal; honours which he enjoyed to a very protracted age.

TIEDEMANN, F. — *Malacology.*

Anatomie de l'Holothuria, &c. Landshut, 1805. folio. One of the finest monographs, according to Cuvier, of the invertebrate animals.

* I consider the arguments hitherto brought forward against Mr. Thompson's views, as particularly weak; and the facts by which they are supported, partial and inconclusive.

TILESIUS, W. G. — *General Natural History.*

A German naturalist and botanist, in the service of
the emperor of Russia : he accompanied captain Krusenstern in his voyage round the world.   His writings
are accurate, but diffuse, and sometimes obscure.   We
have quoted them in regard to some curious fish of
the North Seas, described in the Petersburgh Transactions.

    Annuaire d'Histoire Naturelle.   Leipsig, 1802.
12mo.

TREITSCHKE. — *Entomology.*

Already mentioned as the author of the four last
volumes of Ochsenheimer.

TREMBLEY, ABRAHAM.

Abraham Trembley was born at Geneva, in the year
1700.   His parents were perfectly respectable, although
not in affluent circumstances: they had the means, however, of giving their son the advantage of a good education ;
and he distinguished himself, while at school, in the
study of mathematics.   He declined entering the church,
which his father was desirous that he should do ; but
resolved to travel, and look out for such employment as
suited his attainments.   In this he was fortunate ; for
meeting with lord Bentinck at the Hague, he employed
him as instructor to his children.   It was when residing
in the country with his pupils, that he first noticed the
freshwater polype.   It had previously been observed,
as well as figured, both by Leeuenhoek and Jussieu; but
nothing was known of its history.   By a series of careful observations, continued for a period of nearly four
years, he succeeded in discovering its nature, structure,

mode of nutrition, and propagation; all which are so singular, that, when the account was published, it excited the wonder and admiration of every intelligent mind. Rightly judging that descriptions of phenomena so remarkable could not be well understood without illustrative drawings,,and being unable to execute these himself, he had the good fortune to secure the skilful pencil and discriminating eye of Lyonnet; so that his work was laid before the public in a form suited to the interest and importance of the subject. It was reprinted the same year in Paris; and again in Germany. To mark their sense of the author's merits, the Academy of Sciences elected him a corresponding member, and the Royal Society of London also admitted him into their body. He accompanied his patron to London; and, after a time, made a tour through part of Europe, as tutor to the duke of Richmond. He returned to Geneva in 1757, where he married, and continued to reside for the rest of his life. He died on the 12th of May, 1784. After settling at Geneva, he published two or three works of a religious nature, intended chiefly for the use of young people. A memoir of his life was published at Neufchâtel in 1787, 8vo.; and Sennebier has written his éloge in the " Hist. Lit. de Génève." — J. D.

Mémoires pour servir à l'Histoire des Polypes d'Eau douce, à Bras en forme de Cornes. Leyden, 1744. 4to. pl. 15.

Treutler, F. A. — *Anatomy.*

Observationes Pathologico-Anatomicæ, Auctarium ad Helminthologiam Humani Corporis, continentes. Leipsig, 1798. 4to.

Turton, Dr. — *Conchology.*

A physician, who almost relinquished his profession, from his desire of investigating British conchology.

Besides his writings on this subject, he translated
Gmelin's edition of Linnæus into English.

1. *A Conchological Dictionary* of the British Islands.   Lond. 1819.   12mo. pp. 272. pl. 28.   The
text is according to the Linnæan classification, and
is arranged alphabetically.   The figures are merely
rude outlines, scarcely recognisable.

2. *The British Fauna,* or Compendium of the
Zoology of the British Islands, arranged according to
the Linnæan System.   Swansea. 12mo.   The second
volume, which was to contain the insects, &c., has
never been published.   The type is remarkably
small.

3. *Conchylia Insularum Britannicarum;* or, The
Shells of the British Islands systematically arranged.
Exeter, 1822. 4to. pp. 279. pl. 20.   The divisions
are new ; the genera, those of Lamarck ; some new
are proposed, and characters assigned to them, most of
which are now generally adopted.   The plates are very
accurately designed and engraved by Curtis; but, in
the majority of copies we have seen, they are rather
too highly coloured.   The second volume, intended
to contain the univalves and barnacles, we have not
seen.   Dr. Turton is also the author of another work
on the land and freshwater shells of Britain, which
we have not seen.

VAILLANT, FRANÇOIS LE.—*Ornithology.*

The life and fortune of this celebrated traveller were
devoted to science.   He was born in Surinam, of French
parents ; and, at a very early age, evinced the greatest
devotion to ornithology ; a passion which clung to him
ever after.   Accustomed to investigate nature in the
wild recesses of his native forests, he seems to have
imbibed a strong prejudice against the methods and
systems then in use ; and which no doubt resulted
from the errors, and the artificial modes, of arranging

those beings which he was accustomed to contemplate
in their native haunts. This prejudice, which he never
overcame, has been unfortunate ; as his fame will never
be justly appreciated by those who look upon system and
method as every thing, and the creation of learned names,
and artificial genera, as the ultimate object of science.
To say that Le Vaillant was the first who investigated the
ornithology of Southern Africa, and described more than
500 new species, is saying but little. His divisions of
tribes, founded upon their habits and manners, laid the
foundation for nearly all the new genera of African
birds, since proposed by the great zoological reformers
of the present day. They have, in fact, merely given
Latin names to those which he distinguished by French
ones. In this respect, Le Vaillant has been unjust to
himself, from his overstrained repugnance to system.
He never felt perfectly satisfied, until he had procured
the nest and eggs, and ascertained the sexes, of each
species. He accurately observed their peculiar habits,
and the nature of their food : hence his great work on
the birds of Africa will long remain a source of original
information ; and is as much prized and consulted now,
as it was on its first publication. Systems and methods
change, but nature is always the same. Forgetful that
fortune had not bestowed upon him a princely revenue,
Le Vaillant was inspired with a wish of calling in the
greatest powers of art to illustrate his charming science.
In his *Oiseaux d'Afrique* he was not fortunate in the
choice of his artists: the plates, indeed, have great merit ;
but there does not appear to have been, at that time, a
zoological painter in France of any eminence. The
rising genius of Barraband was soon, however, employed
to decorate his subsequent works ; and the inimitable
plates they contain, will remain a lasting honour to the
talents of the artist, and to the discrimination of his
employer. Of these, it is, indeed, impossible to speak
in terms of too much praise. During his travels in
Africa, Le Vaillant seems to have been patronised by
M. Temminck, the father of the present celebrated orni-

thologist of that name; and the best answer that can be given to those idle stories which have been circulated about his never having been in the countries he described, is, that the birds he shot there are actually in the public Museum of Amsterdam. It could not be expected that publications brought out upon such a magnificent scale, would reimburse their author; still less that they would become a source of profit. Le Vaillant, unfortunately, experienced the truth of the first; yet still, so unabated was his zeal, that, while his patrimony was annually diminishing, he was still projecting works which should, if possible, exceed those which he had already accomplished. Of his private life, the writer can learn but little. From a wish expressed at the conclusion of one of his works (*Ois. d'Af.* vol. vi.), that his sons would complete the remaining portion, it would appear that he was married early; and we have been informed that he united himself again, at rather an advanced age, to a young and amiable woman. During the latter years of his life, his circumstances were unfortunately rather straightened; yet this did not affect his fine flow of spirits, his passion for birds, or his habitual contentment. To a friend of ours*, who visited him in the more *aërial* apartments of a house in Paris, he jocosely observed, " The longer I live, the higher I rise in the world." We should have been glad to have recorded, for the honour of his nation, that this memorable man had passed the evening of his days without feeling the pressure of fortune. His death is said to have been recent.

1. *Voyage dans l'Intérieure de l'Afrique*, par le Cap de Bonne Espérance. Paris, 1790. 2 vols. 8vo.

2. *Second Voyage* dans l'Intérieur de l'Afrique, &c. Paris, 1795. 3 vols. 8vo. Of both these works there are English translations.

3. *Oiseaux de l'Afrique*, Histoire Naturelle des.

---

* The late Dr. Leach.

Paris, 1799—1808. 6 vols. 4to. Each volume contains about 50 coloured plates.

4. *Perroquets,* Histoire Naturelle des.   Paris, 1801. 2 vols. folio. Of this, and the preceding work there are editions in large 4to.

5. Hist. Nat. d'une Partie d'Oiseaux de l'Amérique et des Indes. Paris, 1801. 4to.

6. *Oiseaux de Paradis,* Hist. Naturelle des, et des Rolliers, suivie de celle des Toucans et des Barbus. Paris, 1806. 2 vols. folio.

7. *Promerops et des Guêpiers,* Histoire Naturelle des. Paris, 1807. 1 vol. folio.

VALENCIENNES, A.—*General Zoology.*

The zealous and able continuator of Cuvier's great work on fishes, and author of many Mémoires in the French scientific Transactions.

VALENTYN, F. — *Ichthyology.*

A protestant clergyman, long resident in Amboyna.

The Ancient and Modern History of the East Indies (in Dutch).   Amsterdam, 1724—1726. 5 vols. folio. Cuvier observes, that the third volume contains many details respecting the natural history of the island ; and that the plates of fish are similar to those published by Rénard — (which see).

VANDELLI.—*Ichthyologist.*

A learned Italian naturalist, in charge of the Royal Museum at Lisbon, in the Transactions of which institution he has described several remarkable fishes.

VANDER LINDEN, P. L.—*Entomology.*

A very able writer on the *Hymenoptera;* whose work on the European species presents the first careful elaboration of a portion of the order.

Observations sur les Hyménoptères d'Europe de la Famille des Fouisseurs. 1 vol. 4to. Breuxelles, 1829.

Van Hasselt.—*General Zoology.*

A young and talented zoologist, who was sent to Java with Kuhl, to collect for the Museum of the Netherlands. They unfortunately both perished there, after having made immense collections. Numerous fishes, first discovered and named by them, are inserted in Cuvier's great ichthyological work.

Vièillot, L. P.—*Ornithology.*

An eminent reformer of systematic ornithology, and an indefatigable writer.

1. *Oiseaux Chanteurs* de la Zone Torride, Histoire Naturelle des plus beaux. 1 vol. fol. Paris, 1805.

2. *Oiseaux de l'Amérique Septentrionale,* Histoire Naturelle des. Paris, 1807. folio. Only two volumes of this work have appeared. It contains many valuable observations.

3. *Oiseaux Dorés,* Histoire Naturelle des Colibres, Oiseaux Mouches, Jacamars, et Promerops, aussi des Grimperaux, et des Oiseaux des Paradis. Paris, 1822. This magnificent work owes its chief scientific value to the descriptive portion, which is from the pen of M. Vieillot. The plates are executed by M. Audebert, and, although tolerably accurate, are not very good.

4. *Analyse* d'une nouvelle Ornithologie Elémentaire. Paris, 1816. 8vo. It is generally believed that this pamphlet was written and published to anticipate the labours of Cuvier in this department: it certainly bears evident marks of haste. The genera are intimated with great brevity, and often so imperfectly, that they could not be understood but for the type or example which is quoted for each.

5. *Ornithologie Française,* ou Histoire Naturelle, générale et particulière, des Oiseaux de France. Pàris, 1823. 4to. Published in numbers, each containing six coloured plates.

6. *Galerie des Oiseaux* du Cabinet d'Histoire Naturelle du Jardin du Roi. Paris, 1821—26. 4to. The figures designed by M. Paul Audart: completed, we believe, in 80 numbers. This publication owed its origin, also, to the Planches Coloriées of M. Temminck, to which it is superior in the descriptions, but very inferior as to the execution of the plates.

7. *Histoire Naturelle des Mammifères.* Paris, 1819—22. folio. Each number contains six coloured plates.

Vigors, N. A.—*Ornithology and Entomology.*

An accomplished and philosophic zoologist, many years editor of the Zoological Journal, who was one of the first to apply the circular theory to the arrangement of birds. He is the author of several valuable papers in Transactions and periodicals, but has not published any distinct work on zoology. His paper (in conjunction with Dr. Horsfield), on the birds of Australia, in the *Linnæan Transactions,* has been erroneously attributed by Cuvier to me.

Villers, Charles de.—*Entomology.*

Linnæi Entomologia Fauna Suecica Descriptionibus aucta, Scopoli, Geoffroy, De Geer, Fabricii, Schrank, &c., Speciebus vel in Systemate non enumeratis, vel nuperrime detectis, vel Speciebus Galliæ Australis locupletata. Lugduni, 1789. 4 vols. 8vo. pl. 11. 4to. The author of this compilation considered he was benefiting science, by reducing all the new genera and species of the authors he has enumerated, to the Linnæan nomenclature. In this unconscious effort to stop the course of knowledge, he has only rendered his

work of little or no value to modern entomologists.
The plates, however, are executed with great truth,
and deserve quotation.

VIREY, J. J.— *General Zoology.*

A learned physician and philosophic naturalist of
France, by whose researches was first discovered the
five natural divisions of the animal kingdom.   He was
long editor of one of the principal French medical
journals, in which many of his shorter essays are
contained.

1. *Histoire des Mœurs* et de l'Instinct des Animaux.   Paris, 1822. 2 vols. 8vo.
2. *Philosophie de l'Histoire Naturelle,* ou Phéno-
mènes de l'Organisation des Animaux et des Végétaux.   Paris, 1835. 8vo.

VIVIANI, DOMINICO.— *Zoology and Botany.*

A Catholic ecclesiastic, and Professor of Natural History and Botany at Genoa.

Phosphorescentia Maris, Quatuordecim Lucescen-
tium Animalculorum novis Speciebus illustrata.   Genuæ, 1805. 1 vol. thin 4to.

VOSMAER, ARNOLD.— *Mammalogy.*

Director of the museum and menagerie of the stadtholder.   Died in 1799. He published, both in French
and Dutch, a great number of essays and papers, containing monographs and descriptions of different animals,
the most valuable of which is probably the following:—

Description de l'Espèce de Singe, nommé Orang
Outang, de l'Isle de Borneo.   Amsterdam, 1778.
pp. 23. pl. 2.

WAGLER, J. — *Erpetology. Ornithology.*

A laborious nomenclator and describer of birds and reptiles, and a great projector of new genera. His descriptions are usually very accurate, but are always diffuse; and his writings, destitute of comprehensive views, evince little or no genius.

> *Systema Avium.* 1 vol. 12mo., containing monographs of several genera. He also assisted in editing the discoveries of Spix and Martius in Brazil. (See SPIX.)

WALBAUM, J. J. — *Erpetology.*

Physician at Lubeck. Born in 1724.

> *Chelonographia* (in German). Lubeck et Leipzig, 1782. 4to.

WALCKENAER, C. A. — *Entomology.*

Member of the Academy of Inscriptions and Belles Lettres at Paris.

> 1. *Faune Parisienne,* ou Histoire abrégée des Insectes des Environs de Paris. Paris, 1802. 2 vols. 8vo. pl. 7.
> 2. *Tableaux* des Aranéïdes. Paris, 1805. 1 vol. 8vo. pl.
> 3. *Histoire des Aranéïdes,* published uniformly with Panzer's " Fauna Germanica." A few numbers only have appeared.

WATERHOUSE, GEORGE H. — *General Zoology.*

A most acute and indefatigable zoologist, curator of the Zoological Society's Museum in London. Independent of many valuable contributions in the scientific periodicals and Transactions, he is the author of—

> *The Mammalia* discovered by Darwin during the

voyage of H. M. S. Beagle, in 1832–39, of which 3 parts are published, in 4to. with beautiful plates.

WATERTON, CHARLES. — *Traveller.*

An unscientific, but a very observing naturalist, whose American travels contain many excellent observations on the animals of Guiana and Demerara. An octavo edition was published in 1829.

Wanderings in South America, &c., in the years 1812—14. 1 vol. 4to. London, 1825.

WEBER, F. — *Entomology.*

A German entomologist; Professor at Kiel.
Observationes Entomologicæ. Kiel, 1801. 8vo.

WESTWOOD, J. O. — *Entomology.*

Secretary to the Entomological Society of London. A laborious entomologist, and author of numerous papers in the Transactions and periodicals, many of which are of much value. He has also edited an edition of Drury (which see).

1. *Entomologist's Text-Book.* Foolscap 8vo. London, 1838. A popular compendium.

2. *Introduction* to the Modern Classification of Insects. With numerous woodcuts. 2 vols. 8vo. London, 1839. A valuable work of reference : more calculated, however, for the scientific entomologist, than as an introduction; it seems to us by far too technical and undigested for the student.

WHITE, THE REV. GILBERT. — *Natural History.*

An accurate observer of nature, whose writings, from their agreeable popularity, are in much repute. Numerous editions have been printed of

The Natural History and Antiquities of Selborne, in the county of Southampton. 1 vol. 4to. 1813. Subsequent editions have appeared in octavo.

WIEDEMAN, G. R. C. — *Entomology.*

One of the first authorities upon dipterous insects, of which his descriptions are models of accuracy. It is to be regretted he has published no systematic views on their arrangement, as his principal work is merely supplementary to that of Meigen. His chief works are—

1. *Nova Dipterorum* Genera, Iconibus illustrata. 1 vol. 4to. Kiliæ, 1820.

2. *Diptera Exotica*, 8vo. Pars prima. Kiliæ, 1821. with 2 plates.

3. *Analecta Entomologia* (Diptera, and six other Insects), 4to. Kiliæ, 1824. 1 plate.

4. *Ausser Europäische* Zweiflengelige Insecten (the Diptera of Europe), with 12 plates. 2 vols. 8vo. 1828—30.

He also has edited a valuable periodical entitled *Zoologisches Magazin*, 2 vols. 8vo. Kiel and Altona, 1817 —1823, in which are many of his other papers on foreign Diptera.

WILLUGHBY, FRANCIS. — *General Zoology.*

Francis Willughby, at first the pupil, then the friend and patron, of Ray, and the most able zoologist of his time, was born in Lincolnshire, in the year 1635. His family was of wealth and influence, connected both by the father and mother's side with noble houses of the same name, but in different parts of the kingdom. He studied at Trinity College, Cambridge, under the tuition of Ray, who was his senior only by about seven years. We are not aware whether his mind was first led to the study of nature by associating with his amiable preceptor, but the connection was the means of fostering his attachment to it. It appears from Dr. Derham's

statement, that Willughby was an accomplished botanist,
as could scarcely fail to be the case under such tuition;
but that department he soon left, in a great measure, to
Ray, and devoted himself to other branches of natural
history, particularly zoology. By mutual co-operation,
they seem to have entertained the design of working out
a complete system of nature, of which Derham gives the
following account : —

" These two gentlemen, finding the history of nature
very imperfect, had agreed between themselves, before
their travels beyond sea, to reduce the several tribes of
things to a method ; and to give accurate descriptions of
the several species, from a strict view of them. The
province Willughby had taken for his task," he afterwards states, " was animals (that is, birds, beasts, fishes,
and insects), as Mr. Ray had that of plants. And in
these matters he was a great master, as he was also in
plants, fossils, and, in short, the whole history of nature.
And in the pursuit and acquirement of this knowledge,
he stuck neither at any labour or cost."

The various journeys undertaken with a view to examine the natural history of different countries, already
alluded to in the life of Ray, must obviously have been
defrayed by the liberal hand of Willughby. They
visited, in company, most parts of Europe then resorted
to by British travellers, and lost no opportunity of adding to their knowledge or increasing their collections.
Some time after his return from his foreign tour, Willughby married, in 1668, the daughter of sir Henry
Bernard, and settled at his family residence, Middleton
Hall, in Warwickshire. Here he continued to prosecute
his studies in zoology with the utmost zeal, aided occasionally by his former tutor, who was his frequent
visitant. We are not aware that Willughby ever published any thing during this period, except a paper in
the *Philosophical Transactions* (anno 1671, No. 76.),
entitled, " Observations on a Species of Wasp called
Ichneumon;" for he appeared too absorbed in his great
work to publish the materials in detached papers. Eight

years subsequent to his marriage were spent in the manner indicated, during which he had accumulated a great mass of materials on zoology; but all his exertions were cut short by an attack of illness, to which he fell a victim on the 3d of July, 1672, in the thirty-seventh year of his age. He left two sons, Francis and Thomas, who were intrusted to the charge of Ray, to whom he bequeathed an annuity of 60*l.* as a further mark of his regard. The former of these youths died young; the other was created a peer by queen Anne, with the title of lord Middleton.

The first of his works that appeared, under the editorial superintendence of Ray, was entitled " Francisci Willughbeii Arm. Ornithologiæ Libri Tres; in quibus Aves omnes hactenus cognitæ, in Methodum, Naturis suis convenienter reductæ, accuratè describuntur. Descriptiones elegantissimis auri incisis illustrantur. Totum Opus recognovit, digessit, supplevit Johannes Raius." Lond. fol. This work was printed at the expense of the author's widow. It was followed by another on fishes, printed at the expense of the Royal Society. In this instance the editor may almost be regarded as co-author; for, besides revising and arranging the whole, he added the whole of the first and second books. It is clear, however, that this is essentially the result of Willughby's labour, and fully entitles him to be called the father of systematic zoology in this country. We have endeavoured to form an estimate of his merits, when speaking of the influence which his works have exercised on the progress of natural history.* — J. D.

1. *Willughbeii Ornithologiæ,* Libri Tres : totum re-cognovit, digessit, supplevit, Joan. Raius. Lond. 1676. folio. pp. 307. pl. 77.

2. *The Ornithology of Francis Willughby,* of Middleton, in the County of Warwick, Esq., F.R.S. London, 1678. folio. pp. 441. pl. 78.

* See also Preliminary Discourse, p. 25.

WILSON, ALEXANDER. — *The Ornithologist of America.*

 England, no less than America, has reason to be
proud of the name of Wilson. He was born of humble
though respectable parents, at Paisley, about the year
1768, and received the elements of a classical education
at the grammar school of his native town. His own
mother died when he was but ten years old ; and this
misfortune was rendered doubly distressing to him, as
he soon after incurred the persecution of a stepmother ;
to such an extent was this carried, that the poor lad
was compelled to forsake his home, and seek an asylum
with a relation. With Mr. Duncan, a weaver, he
found kindness and hospitality ; and under his protection
young Wilson set about with diligence to acquire a
knowledge of the same trade, in which he continued
for several years. Even at this early period of his life,
he studiously devoted every interval of leisure to the
acquirement of knowledge. He read whatever books
came in his way ; and from a perusal of Burns, imbibed
a portion of the poetic fire of that son of genius. Meantime Mr. Duncan relinquished his trade of weaving,
and became a travelling pedlar : he was joined in this
humble occupation by young Wilson, who thus became
tinctured with that love for travelling, which subsequently grew into a ruling passion. His new mode of
life was certainly favourable to his poetic genius : he had
nature before him, and while journeying through the
wild and romantic scenes of his native country, he no doubt
composed the greatest part of the " Poems, humorous,
satirical, and serious," which came out under his name,
when he was only twenty-two : they went through two
editions, although the author does not appear to have
derived profit by their success. About this time he took
an active part in the disputes between the master and
operative weavers of his native town ; defending the
cause of the latter by several spirited letters, and by a
keen satire, addressed, in the Scottish dialect, to the most

active of their adversaries. This piece was, however, personal, and, however justly applied, the law considered it as a libel, and the young champion was sentenced to a short imprisonment. In 1792, he published " Watty and Meg," anonymously; and such were its superior merits, above any thing he had before written, that it was by many attributed to Burns. Like that unfortunate genius, Wilson had to contend against poverty. All admired his poems, and the judicious few discerned the indication of greater powers, which only wanted some fostering hand to protect and bring forward. But that hand was never extended. He was poor, and soon became dissatisfied. He saw no prospect of bettering his condition in his native country, and resolved to seek his fortune on a foreign shore. He landed in America in 1794, without a friend in the world, and with only a few shillings in his pocket. His courage, however, did not forsake him; he threw his wallet on his back, and his gun on his shoulder, and set off on foot for Philadelphia. Here he soon found employment, and returned to his old trades; first as a weaver, and subsequently as a pedlar. These humble occupations, however, were ill suited to his talents ; and, in the following year, we find him a schoolmaster, " studying mathematics with great diligence and success." Whether the natural restlessness of his disposition, or local causes of discontent, induced him to these frequent changes, certain it is, that he never remained long in any one place ; as he appears to have removed to three or four towns in a few years, teaching at their schools, or conducting one of his own. During one of these intervals of employment, he made a journey on foot into the Genesee country ; traversing an extent of near 800 miles in twenty-eight days. At one of these stations he formed an intimacy with the celebrated naturalist, Bartram; and from this period we may probably date the commencement of that love for natural history, which soon grew into a ruling passion. Enjoying the society and friendship of such

a man, with the advantages of his library, and the benefits of his knowledge, Wilson soon made rapid progress in his ornithological pursuits.

Yet still he had to struggle against poverty ; the
scanty remuneration arising from his school was barely
sufficient to supply the necessaries of life, and his spirits
sank under feelings of melancholy and despondence:
these were probably increased by devoting those short
intervals of leisure he was able to snatch from his se-
dentary occupation, to poetry and music, instead of
seeking to divert and exhilarate his spirits by exercise.
Under the hospitable roof of Mr. Bartram, he had made
many friends, who perceived this morbid sensibility with
much alarm.   One of these, to whom he communicated
his feelings, persuaded him to divert his mind by a new
pursuit; and recommended drawing.   The expedient
was a happy one ; Wilson adopted it with eagerness,
and soon made rapid progress in this delightful art.   In
a simple and modest note, addressed to Mr. Bartram
soon after his first commencement, and sent with some
of his sketches, he says, " The duties of my profession
will not admit me to apply to this study with the assiduity and perseverance I could wish : the chief part of
what I do is sketched by candle-light; and for this I am
obliged to sacrifice the pleasures of social life, and the
agreeable moments I might enjoy in company with you
and your amiable friend."   In 1804, he first contemplated his great work, and soon communicated his wishes
to his friends : they approved of the design, but saw
plainly, that, in his present circumstances, it was too
vast an undertaking.   This opinion did not, however,
discourage him ; and he employed his next vacation in
a journey to the Falls of Niagara,—" earnestly bent," as
he himself says, " on making a collection of all the
North American birds."

He set out on foot ; and in 59 days travelled nearly
1200 miles, 47 of which he performed on the day he
returned home.   The pleasures of this journey appear to have awakened all his old feeling about tra-

velling ; and his love for ornithology seems to have
dilated into a sudden and passionate desire of acquiring
a general knowledge of natural history, and then of
wandering over unexplored regions. In this new scheme,
Wilson, like too many others of the same ardent temperament, appears to have forgotten the pecuniary means
by which his object was to be attained ; for, at this very
time, he told a friend, he was not possessed of more
than four shillings !   The scheme, however, seems to
have taken possession of his mind, and he only waited
for a fit opportunity to carry it into execution.   In
the meantime, he devoted all his leisure to ornithological pursuits, executing drawings of native birds, and,
with little or no assistance, producing, in a short time,
some spirited plates, etched upon copper.   After at-
tempting in vain to induce his friend, Mr. Lawson, the
engraver, to join him in an American ornithology, and
having now satisfied himself of his ability to undertake
the plates, he solemnly declared he would prosecute it
alone, even at the hazard of his life.   " I shall at least
leave," added he, " a beacon to point out where I perished."   Against such enthusiasm, seconded by innate
talents and persevering industry, no obstacles can finally
stand ; but an incident about this time occurred, which
endangered the execution of this mighty project.

Encouraged by his valuable friend Mr. Bartram, he
made an offer of his services (conveyed in a sensible,
modest, and well-expressed letter to the president, Jefferson) to accompany the expedition under major Pike,
then fitting out by the government for exploring the
banks of the Missouri, &c.   Yet with abundant proof
before him of the peculiar talents of Wilson, and of the
great advantage that science would have derived from
the services of such a naturalist, it must be recorded,
to the disgrace of Jefferson, that, upon this occasion, he
neither supported the character of a statesman or a
gentleman : he neither accepted the offer of Wilson, nor
had the courtesy to return him an answer.   The liberal
spirit swells with indignation at such treatment ; but

similar instances on this side the Atlantic are still fresh
in the memory of those who will hand them down to
posterity, with the stigma of having blasted the hopes
of rising genius, and thwarted the advancement of intellectual knowledge. It has been said, that England
can rarely be accused of injustice to the talents of
her sons. I do not think so. In those classic haunts
that nurtured the piety and the philosophy of a Ray
and a Lister, their names are scarcely known ; and
their memory has only now been recalled to our recollection by the eulogiums of a distinguished foreigner.
Wilson received that assistance and support from liberal
individuals in a foreign land, which was withheld from
him in his own: while the ardent and intrepid Bowdich sought refuge from the persecutions of a court
party, and the calumnies of the Quarterly Review, in
the friendship and protection of a rival nation.

Mortified as Wilson must have been at the cold
and contemptuous neglect which he experienced on
this occasion, the time was now approaching, when his
talents were to raise him from the penury and obscurity against which he had so long struggled. Mr.
Bradford, an eminent bookseller at Philadelphia, undertook to publish an American edition of Rees's Cyclopædia; and was so pleased with Wilson, on his first
introduction, that he offered him the business of assistant editor, with a liberal salary, which was of course
joyfully accepted. From this time we hear no more of
his duties as a wandering schoolmaster; but find him
applying with such diligence to his new occupation, that
in a short time his health became considerably impaired.
To recruit both his mind and body, he set out on one
of his pedestrian journeys; having first attained that
object which had long been nearest his heart, namely,
inducing Mr. Bradford to advance the necessary funds
for the publication of his *American Ornithology*. This,
and all his former journeys, were made subservient to
this great object; and on his return, he devoted himself
to it, at every moment of leisure, with so much perse-

verance, that at the close of the following year (1808)
the first volume was ushered into the world.

The publication of such a book, — which would have
done honour to the presses of Europe, — in such an
infant republic as America, excited the greatest astonishment.   The admiration it produced was universal; and
so completely did this praise confirm Wilson in his resolution to continue it, that in the very same month he
actually set off on foot, upon " a journey to the east-
ward, to exhibit his book, and procure subscriptions."
His reception, as might naturally be expected, was various; but upon the whole, his success, during this and
a subsequent journey into Georgia, was sufficiently great
to call for a new edition of the first volume, which increased the number of copies to 500.   The second volume was published early in 1810; and immediately
after, this extraordinary man set out alone, in a small
boat, to descend the Ohio, a distance of 700 miles!
From Louisville he pursued his journey on foot for
seventy-two miles; and then, purchasing a horse, continued his route to Natchez alone, where he safely arrived, after traversing 678 miles.   Wilson was not a
man to be daunted by ordinary hardships, but those he
must have experienced on this occasion were very great.
Alone, encumbered with his baggage, cutting his way
through dreary cane swamps and morasses, and narrowly
escaping death from a sudden illness, while exploring
these dreary solitudes, Wilson might truly say, " I
have gone through difficulties that few can have a conception of."   During this journey, he appears to have
added many names to his list of subscribers, and acquired valuable materials for his work.   On many occasions, his reception among the inhabitants was marked
by unusual hospitality.   One instance we cannot forbear
quoting: — " My hospitable landlord refused to take
any thing for my fare, or that of my horse, saying, ' You
seem to be travelling for the good of the world; and I
cannot, I will not, charge you any thing.   Whenever
you come this way, call and stay with me, and you shall

be welcome.'" This man's name was Isaac Walton — a
name immortalised by " The Complete Angler," and
synonymous with every thing that is kind and benevolent. On his return, he took up his residence at the
Botanic Garden of his friend, Mr. Bartram, where, free
from interruption, and surrounded by agreeable objects,
he made rapid progress in his work. In the beginning
of 1813, the seventh volume was published ; but the
near prospect of a termination to his labours, instead
of lessening his application, tended, unfortunately, to
increase it. That he might more readily superintend
the mechanical department of printing and colouring, he
quitted the quiet retreat he had chosen, and removed
into town. Here it was, that, deprived of his usual exercise, and oppressed with additional labour, his health
became debilitated, and his mind harassed. A severe
attack of dysentery ensued, and a few days' illness closed
the life of Wilson.

The merits of this extraordinary man, as a naturalist,
are of a peculiar order. He was no systematist; and it
is not surprising — when we consider the then state of
ornithological science — that we should find him occasionally expressing his contempt for the systems then in
use. These early prejudices against systematic classification are seldom overcome : they were, indeed, not
so strong in Wilson as in Le Vaillant; but in both cases
they seem to have operated, with other causes, in preventing these eminent men from paying little or no attention to this part of the science. Wilson is a describer
of nature, not as she appears through the medium of
books, or is exhibited in the glazed cases of museums.
He sought her in her native wilds, and his descriptions
seem penned upon the spot. He has written with all
the truth and accuracy of a naturalist, and all the
warmth and delightful enthusiasm of a poet ; his descriptions, in fact, are biographies; and cold indeed must
be that heart which cannot participate in the feelings of
the writer. The little attention he bestowed on system
is not to be regretted ; for, by studying nature in her

animated state, he has done more real service to science than all who have preceded or followed him in the same tract. Fascinating indeed would be the study of ornithology, if we could hope to see the birds of Europe described and figured by the pen and pencil of a Wilson. America may well feel proud at having fostered such a man; for we question very much, if, with all his abilities, he would have received half as much encouragement in Great Britain. In his general character, Wilson possessed all those virtues and failings common to an enthusiastic temperament. A nice sense of honour led him, some-times, to be "sudden and quick in quarrel;" but it was the burst of the moment, and his temper soon regained its wonted placability, kind-heartedness, and benevolence. Intent upon improvement, his industry and perseverance were most extraordinary. " It ever gave him pleasure to acknowledge error, when the conviction resulted from his own judgment alone; but he could not endure to be told of his mistakes." Finally, he was remarkably tem-perate, scrupulously just, and had the greatest veneration for truth.

Science is indebted to Mr. Ord, the friend and asso-ciate of Wilson, for the publication of the ninth volume of the American Ornithology, and for an interesting ac-count of his life, from which this sketch has in a great measure been taken. We are rejoiced to hear that a gentleman, said to be well qualified for the task, has, published or is about to publish, a new edition of this na-tional work at Philadelphia.

American Ornithology, or the Natural History of the Birds of the United States; illustrated with Plates. By Alexander Wilson. Philad. 1808—14. 9 vols. thin folio, plates 76. The best English edition is that edited by sir W. Jardine.

WILSON, JAMES. — *General Zoology.*

An accomplished and elegant writer on zoology, who has contributed largely to the current edition of the *Encyclopædia Britannica.*

Illustrations of Natural History, published in numbers, at Edinburgh, in folio, with well-engraved coloured plates.

WOLFF AND MEYER. — *Ornithology.*

Tasschenbuch der Deutschen Vögelkunde (Register of the Birds of Germany), 2 vols. 8vo. Franc. 1810. The first volume contains the land birds by Dr. Wolff; the second, those of the water, by Dr. Meyer. Cuvier remarks: — " Cet ouvrage est plein de très-bonnes observations." The same authors seem to have published another work in German, which we have not seen, entitled

Naturgeschichte der Vögel Deutschlands.

WOOD, WILLIAM. — *Conchology.*

A Fellow of the Royal Society, and formerly in the medical profession ; but now the most learned bookseller in London for works connected with natural history. Although Mr. Wood is a strict Linnæan, his descriptions are accurate and very useful.

1. *General Conchology,* or a Description of Shells, arranged according to the Linnæan System. Lond. 1815. 1 vol. 8vo. pp. 246. pl. col. 60. The plates are very good, being drawn and etched by Sowerby. This, like all other General Conchologies attempted in this country, was discontinued after the first volume, from want of support.

2. *Zoography,* or the Beauties of Nature displayed, in select Descriptions from the Animal, Vegetable, and Mineral Kingdoms ; systematically arranged. 3 vols. 8vo. With picturesque plates by Daniel.

3. *Index Testaceologicus,* or a Catalogue of Shells, British and Foreign, arranged according to the Linnæan System, with Latin and English Names, References to Figures, and Places where found. London, 1818. With miniature figures of numerous species.

These figures are executed with great neatness, and often with beauty. The arrangement, however, is that of Linnæus, and the synonyms often short and incorrect. Mr. Wood, jun., by whom these plates were executed, has used every endeavour to procure original specimens for delineation; but when the species could not be found in the London cabinets, he has very properly copied the original figure, quoted by Linnæus, or by his followers. A Supplement has since been added.

Wormius, Olaus.

Professor at Copenhagen. Born in 1588; died in 1654.

Museum Wormianum. Leyden, 1655. 1 vol. folio.

Yarrell, William. — *Ichthyology and Comparative Anatomy.*

One of the most eminent ichthyologists in this country, and a skilful comparative anatomist. Independent of his papers in our scientific journals, the Linnæan and Zoological Transactions, &c., he has published the following: —

1. *The History of British Fishes.* 2 vols. 8vo. and a Supplement. A most beautiful and valuable work; the only one, in fact, upon this department of our native fauna, besides that of Pennant. The woodcuts of fish, and the landscape vignettes, are of equal merit with the letterpress.

2. *British Birds*, the Natural History of, by the same author, is now publishing in parts.

Zeder, J. D. H. — *Intestinal Worms.*

1. *Natural History* of the Intestinal Worms of Goeze (in German). Leipsig, 1800. 1 vol. 4to.

2. *An Introduction* to the Natural History of

Intestinal Worms (in German).   Bamberg, 1803.
1 vol. 8vo.

ZETTERSTEDT, J. G. — *Entomology.*

A Swedish entomologist, who has written on the
insects of his own country.

1. *Orthoptera Sueciæ.* Lundæ, 1811. 1 vol. 8vo.
2. *Fauna Insectorum Lapponica.* Part I. Ham-
monæ, 1826. 1 vol. 8vo. Containing the coleoptera,
orthoptera, and hemiptera.

ZIMMERMAN, E. A. W. — *Mammalogy.*

1. *Specimen Zoologiæ* Geographicæ Quadrupedum,
Domicilia et Migrationes sistens, dedit Tabulamque
Mundi Zoographicam. Lugd. Bat. 1777. 4to.
pp. 686. pl. 1.
2. *Déscription d'un Embryon d'Eléphant,* accompagnée de quelques nouvelles Observations sur l'Histoire Nat. de ce Quadrupède. Erlang, 1783. 4to.
pp. 20. pl. 1.

ZINKE, GEO. GOTTFR. — *Entomology.*

Naturgeschichte, &c.   Histoire Naturelle des Insectes invisible aux Arbres verts, et Moyen de les
détruire. 1 vol. 8vo.   Weimar, 1798.

ZSCHUCHII, J. J. — *Entomology.*

Museum Leskeanum, pars Entomologica.   With
3 plates. 2 vols. 8vo.   Lipsiæ, 1788.

APPENDIX.

The following names have been accidentally omitted in the regular alphabetical series.

BURCHELL, DR. WILLIAM J. — *The African Traveller.*

ONE of the most learned and accomplished travellers of any age or country, — whether we regard the extent of his acquirements in every branch of physical science, or the range of the countries he has explored. Science will ever regret that one whose powers of mind are so varied, and so universally acknowledged throughout Europe, should have been so signally neglected by his government, — the most thankless and ungrateful one, to unpatronised talent, under Heaven. Having expended large sums in prosecuting his travels in Southern Africa, and bringing home immense collections, astronomical observations, &c., the Prussian government offered him a handsome pension, if he would carry all to Berlin, and settle in that city. This he refused, under the vain hope of publishing his discoveries in his own country. Disappointed in this, he again set off for Tropical America, where he travelled for nearly seven years. The fruits of all these labours, however, lie hid in unopened packages, and may probably never see the light until the death of their possessor. A government which bestows honours upon writers of novels, and pensions for licentious ballads, cannot be expected to regard modest worth or unobtrusive talent.

Travels in the Interior of Southern Africa. 2 vols. 4to. 1822—24.

EYTON, J. C.—*Ornithology*.

1. *British Birds.* A History of the rarer British Birds; with Woodcuts. London, 1836. 1 thin vol. 8vo.

2. *Monograph* of the Anatidæ, or Duck Tribe.* London, 1838. 1 vol. 4to. with numerous plates, chiefly anatomical, and very well executed. But the nomenclature, both of this and the former work, is often erroneous; the author, moreover, seems not to have sufficiently studied the previous labours of others.

GOULD, JOHN.—*Ornithology*.

A zealous and very able ornithologist, now travelling in Australia, who has published some valuable, although very expensive, works upon birds; the chief of these are —

1. *Century* of Birds from the Himalaya Mountains. 1 vol. folio. London, 1832—3.

2. *Birds of Europe.* 16 parts, at three guineas each.

3. *Ramphastidæ*, Monograph of the. 3 parts.

4. *Trogonidæ.* Ditto.

We trust the author will hereafter reprint these expensive volumes in such a form as that they may be accessible to naturalists; and thereby diffuse science, instead of restricting it to those only who are wealthy.

---

* The writer, in his preface, informs the public, that "the greater part" of my account of the duck tribe in the *Menagerie of Animals* "is copied from Latham." I will now put it to himself, whether this assertion is true or false. My account of this tribe occupies eighty-two pages and a half (that is, from 190. to 278.); while that portion, which is thus truly and avowedly copied from Latham, fills exactly five pages and a half. The *reason* of this being copied, is expressly stated; it is, to give the Doctor's own words for species that nobody else has seen or heard of.. I know not who Mr. Eyton may be; but I will tell him, as a friend, that when once an author is detected in a dishonest statement, he loses the confidence of his readers; and that the depreciation he hopes to effect towards others, will assuredly fall upon himself. After this exposure, the public will judge how far the additional assertion of there being "little new" in what I have written, is to be believed. Even now, where one reader peruses his work, twenty will peruse this; the sale of each being nearly in this proportion.

HAWORTH, ADRIAN H.—*Entomology and Botany.*

A distinguished British entomologist, whose writings on the *Lepidoptera* will be always valuable. He possessed a most extensive collection of insects, and has largely contributed to systematic botany, by many valuable publications on the ice-plants, the African aloes, &c. He died, much respected, at Chelsea, in 1833.

Lepidoptera Britannica, sistens Digestionem novam Insectorum Lepidopterorum quæ in Magna Britannia reperiuntur. 1 vol. 8vo. London, 1803—1828. The number of species enumerated amount to 1450.

MACGILLIVRAY, W.—*Ornithology.*

1. *Rapacious Birds* of Great Britain. 1 vol. 12mo.
2. *British Birds.* 3 vols. 8vo. We are unacquainted with either of these works.

NEWMAN, EDWARD.—*Entomology.*

Besides several descriptive papers in periodicals, he has published—
　1. *Sphinx Vespiformis,* &c. London. One thin volume, 8vo. A work already alluded to.*
　2. *Grammar* of Entomology. 1 vol. 18mo. London.

OWEN, RICHARD.—*Zoological Anatomy.*

The distinguished Professor of Anatomy at King's College, and curator of the museum of the College of Surgeons. His numerous and elaborate papers have been hitherto scattered in transactions and periodicals.

* Geography and Classification of Animals, p. 220.

WALKER, FRANCIS. — *Entomology.*

A laborious entomologist, well known for his almost exclusive study of the *Chalcididæ.* Nearly all his papers are inserted in the

Entomological Magazine, of which he was the chief editor. London, 1832—38. 8vo.

INDEX.

PART I.—TAXIDERMY.

PART II.—BIBLIOGRAPHY AND BIOGRAPHY.

THE END.

LONDON:
Printed by A. SPOTTISWOODE,
New-street Square.

Printed in the United States
By Bookmasters